Chromatography of Polymers

ACS SYMPOSIUM SERIES **521**

Chromatography of Polymers

Characterization by SEC and FFF

Theodore Provder, EDITOR
The Glidden Company
(Member of ICI Paints)

Developed from a symposium sponsored by the Divisions
of Polymeric Materials: Science and Engineering, Inc.,
and Analytical Chemistry
of the American Chemical Society
at the Fourth Chemical Congress of North America
(202nd National Meeting of the American Chemical Society),
New York, New York,
August 25–30, 1991

American Chemical Society, Washington, DC 1993

Library of Congress Cataloging-in-Publication Data

Chromatography of polymers: characterization by SEC and FFF / Theodore Provder, editor.

p.   cm.—(ACS symposium series, ISSN 0097–6156; 521)

"Developed from a symposium sponsored by the Divisions of Polymeric Materials: Science and Engineering, Inc., and Analytical Chemistry of the American Chemical Society at the Fourth Chemical Congress of North America (202nd National Meeting of the American Chemical Society), New York, New York, August 25–30, 1991."

Includes bibliographical references and index.

ISBN 0–8412–2625–3

1. Polymers—Analysis—Congresses.   2. Field-flow fractionation—Congresses. 3. Gel permeation chromatography—Congresses.

I. Provder, Theodore, 1939–   . II. American Chemical Society. Division of Polymeric Materials: Science and Engineering. III. American Chemical Society. Division of Analytical Chemistry. IV. Chemical Congress of North America (4th: 1991: New York, N.Y.) V. Series.

QD139.P6C49   1993
668.9—dc20   93–186
                                                                           CIP

668.9
C557
1993

Foreword

THE ACS SYMPOSIUM SERIES was first published in 1974 to provide a mechanism for publishing symposia quickly in book form. The purpose of this series is to publish comprehensive books developed from symposia, which are usually "snapshots in time" of the current research being done on a topic, plus some review material on the topic. For this reason, it is necessary that the papers be published as quickly as possible.

Before a symposium-based book is put under contract, the proposed table of contents is reviewed for appropriateness to the topic and for comprehensiveness of the collection. Some papers are excluded at this point, and others are added to round out the scope of the volume. In addition, a draft of each paper is peer-reviewed prior to final acceptance or rejection. This anonymous review process is supervised by the organizer(s) of the symposium, who become the editor(s) of the book. The authors then revise their papers according to the recommendations of both the reviewers and the editors, prepare camera-ready copy, and submit the final papers to the editors, who check that all necessary revisions have been made.

As a rule, only original research papers and original review papers are included in the volumes. Verbatim reproductions of previously published papers are not accepted.

M. Joan Comstock
Series Editor

Contents

SIZE-EXCLUSION CHROMATOGRAPHY:
VISCOMETRY DETECTION

Preface

PRODUCT DEVELOPMENT OF POLYMERIC MATERIALS in the 1990s is not a simple linear process from product design to product performance and market introduction. Many constraints are produced by business and social forces. These forces affect and influence the direction of the product development cycle and make the process highly nonlinear and iterative. Often, product R&D, process scale-up, plant manufacturing, and quality assurance overlap, occur in parallel, and feed back to each other in order to shorten the product development and market introduction cycle. Some of the key constraints affecting this process in the 1990s are the following: safety, health, and the environment; product quality; emphasis on customer needs; improved product–process–customer economics; limited development of new commodity polymer building blocks (monomers); and global competition.

Polymer characterization, particularly chromatographic characterization, facilitates and accelerates the product development and market introduction cycle in the context of these key constraints. Two broad areas of chromatographic analysis that play an essential role in the characterization of polymeric materials are size-exclusion chromatography (SEC) and field-flow fractionation (FFF).

About the Book

This book covers significant advances in the chromatographic characterization of polymers by SEC and FFF and is organized into four sections: (1) field-flow fractionation; (2) size-exclusion chromatography: fundamental considerations; (3) size-exclusion chromatography: viscometry detection; (4) size-exclusion chromatography: high-temperature, ionic, and natural polymer applications.

The first section focuses on current developments in FFF methods, particularly sedimentation (Sd) and flow FFF, as well as thermal (Th) FFF techniques for particle size analysis and the characterization of polymer molecular-weight distribution (MWD). Barman and Giddings apply Sd–FFF to narrow and broad particle size distribution latexes with densities lower and higher than the aqueous carrier fluid and also to resolving poly(methyl methacrylate) particle aggregate cluster sizes from singlets to octets. Arlauskas et al. apply Sd–FFF to unique perfluorocarbon emulsions (artificial blood precursors). Ratanathanawongs and Giddings compare flow-FFF and Sd-FFF for particle size analysis and show the advantage of using flow-FFF for analyzing nanometer-size particles. They also make use of the steric FFF mode for analyzing broad distributions containing both submicrometer- and supermicrometer-sized particles. Myers et al. provide an excellent overview of the principles and practices of Th-FFF as applied to synthetic polymers. Schmipf et al. extend Th-FFF to the analysis of copolymers and show that the thermal diffusion coefficient is a linear function of monomer composition for random copolymers and block copolymers in solution. Lee demonstrates how Th-FFF uniquely can be applied to the determination of the gel content of polymers subjected to electron-beam radiation.

The five-chapter section on fundamental considerations of SEC includes a chapter by Hunkeler et al. that establishes critical conditions (solvent power) for separating polymers independent of molecular weight but dependent on copolymer or blend composition. The chapter by Sanayei, O'Driscoll, and Rudin is a landmark study; they develop and experimentally validate a one-parameter expression for the universal calibration curve in the context of a one-parameter correlation between intrinsic viscosity and molecular weight derived from solution property theory. Cheng and Zhao developed a method to determine quantitatively the specific refractive increment by SEC. Gores and Kilz review and explore methods of multidetection SEC for determining composition for a variety of copolymers. Shetty and Garcia-Rubio utilize SEC and spectroscopic analysis to elucidate quantitatively the end groups in poly(methyl methacrylate).

The five-chapter section on SEC and viscometry detection illustrates the continuing interest in the application of viscometric detection for determining the absolute MWD of polymers and estimates of long-chain branching in polymers. In one chapter, Balke, Mourey, et al. develop a strategy and systematic approach for interpreting multidetector SEC data from concurrently used viscometer, light scattering and refractometer detectors; in another chapter they apply this methodology to high-temperature SEC analysis of recycled plastic waste. Lesec et al. explore the question of quantitative accuracy in viscometry detection, and the chapter by Kuo, Provder, and Koehler explores the use of viscometry detection in several solvents. The chapter by Goldwasser is another landmark study, presenting a new method for determining the absolute number-average molecular weight of copolymers, polymer mixtures, and samples of unknown structures by means of SEC utilizing only viscometry detection.

The last section deals with polymer applications involving high-temperature SEC methods, ionic polymers, and natural polymers. Pang and Rudin use SEC with viscometry and light-scattering detection to assess long chain branching frequency in polyethylene; they find the results are in reasonable coincidence with ^{13}C NMR spectroscopy, the referee method. Lehtinen and Jakosuo-Jansson use high-temperature SEC to determine the MWD of poly(4-methyl-1-pentene). Markovich et al. combine high-temperature SEC with Fourier transform IR detection to characterize simultaneously the chemical composition of ethylene-based polyolefin copolymers as a function of branching concentration and MWD. Wu, Curry, and Senak provide a useful overview of the application of SEC to characterize cationic, nonionic, and anionic vinylpyrrolidone copolymers. The application of SEC viscometry to natural products is demonstrated by Timpa for the characterization of cotton fiber and by Fishman et al. for the structural analysis of aggregated polysaccharides.

The chapters in this book represent current significant developments in chromatographic characterization of polymers by SEC and FFF in both academic and industrial laboratories. I hope that this book will encourage and catalyze further activity in the chromatographic characterization of polymers.

Acknowledgments

I am grateful to the authors for their effective oral and written communications and to the reviewers for their critiques and constructive comments.

THEODORE PROVDER
The Glidden Company
(Member of ICI Paints)
Strongsville, OH 44136

January 11, 1993

FIELD-FLOW FRACTIONATION

Chapter 1

Calibration of a Photosedimentometer Using Sedimentation Field-Flow Fractionation and Gas Chromatography

R. A. Arlauskas, D. R. Burtner, and D. H. Klein

Alliance Pharmaceutical Corporation, 3040 Science Park Road, San Diego, CA 92121

Alliance Pharmaceutical Corp. is conducting research on a series of high density perfluorocarbon emulsions stabilized by egg yolk phospholipid. These emulsions are based on 1-bromo-perfluoro-n-octane (perflubron) and are being developed for several applications including imaging, oxygen and carbon dioxide transport. Particle size distribution was measured by Sedimentation Field-Flow Fractionation (SdFFF), photon correlation spectroscopy and photosedimentation. Since there are no standards available with the appropriate density (~ 1.9 g/cc) and optical properties, calibration of the mass distribution was performed by collecting monosized fractions, determining the mass concentration by GC and comparing this value with the SdFFF detector signal. The calibrated mass distribution was then used to correct for the scattering effects of the photosedimentation instruments used for routine quality control measurements. In this way, the test emulsion was used as its own standard.

The emulsion product contains an emulsion particle population with a median particle size of 0.25 μm and a liposomal population of undetermined size which is well resolved from both the void volume and the emulsion particle distribution in SdFFF.

SdFFF is determined to be a useful tool for particle size determination. In contrast with photosedimentation particle sizing instruments, SdFFF allows straightforward calibration with the test material.

A vital component of emulsion characterization is the determination of particle size distribution. The particle size of an intravenous pharmaceutical emulsion is considered to play an important role in the biocompatability, intravascular dwell, toxicity and physical stability of the resulting formulation. However, the determination of the correct particle size distribution is not always a simple task. It is complicated by factors such as the polydispersity of the distribution, the optical properties of the particle and the absolute size of the particles. Alliance

0097–6156/93/0521–0002$06.00/0

Pharmaceutical Corp's. perflubron dispersions are polydisperse, ranging in size from 0.03 μm to 1.0 μm. Figure 1 is a freeze-fracture transmission electron micrograph of a typical 90% (w/v) perflubron dispersion at a magnification of 50,000x. Figure 1 clearly shows the polydisperse nature of these systems. Two types of particles with different particle size, density, and optical properties are present. The larger of these is the true perflubron emulsion particle distribution, droplets of perflubron covered by a layer of egg yolk phospholipid (EYP). The smaller is a population of liposomes which contain no perflubron and are single or possibly multilamellar vesicles of EYP. These two populations have particles of different sizes and densities. It is these characteristics which are vital to the determination of a correct particle size distribution.

The determination of a correct particle size distribution of an emulsion product is important so that the distribution can be monitored and maintained for the purposes of quality control and stability. Typically, it is necessary to analyze standards with known diameters to ensure accurate particle sizing measurements. There are no commercially available standards with the appropriate physical characteristics, density and optical qualities, which can be used to calibrate the particle sizing instruments used for the perflubron emulsions. In addition to this, the large range of particle sizes of the perflubron emulsion necessitate the determination of a scattering correction factor which will allow for determination of the true particle size distribution. Simply put, large particles scatter light well while small particles do not. This results in a distribution skewed towards the large sized end of the distribution. It is therefore necessary to determine the perflubron mass distribution using an independent analytical instrument, one which is easily calibrated with external standards, and compare this distribution with the optical detector response. The calibrated mass distribution can then be used to correct for both the Mie and Rayleigh scattering exhibited by the perflubron emulsion particles.

Particle Sizing. Measurement of the mean particle size can be accomplished by several methods. Photon correlation spectroscopy or quasi-elastic light scattering is a widely applied technique whereby particle size is determined from particle diffusivity. Diffusivity is calculated from the autocorrelation function of the time dependence of scattered light intensity, due to the Brownian motion of the particles. In the case of an ideal unimodal distribution, the method of cumulants provides reliable results. This gaussian analysis provides a mean diameter and standard deviation. A chi-squared parameter is used to determine the goodness of fit. Bimodal distributions, such as observed in the 90% (w/v) perflubron emulsion, shown in Figure 1, require a more complicated approach. For such distributions an attempt is made to determine the set of exponentially decaying functions which, when added together with the appropriate factors and squared, provide the measured autocorrelation function. Each decaying exponential is related to a discrete particle diameter (1).

Photosedimentation, in a centrifugal field, is also simple and frequently performed. In this technique an external centrifugal field is applied, either a fixed rotational speed or one in which the speed is increased. The particles settle at a rate which varies with their diameter. An optical detector, fixed at 560 nm in the case of the HORIBA instrument, measures the change in light transmission as a function of

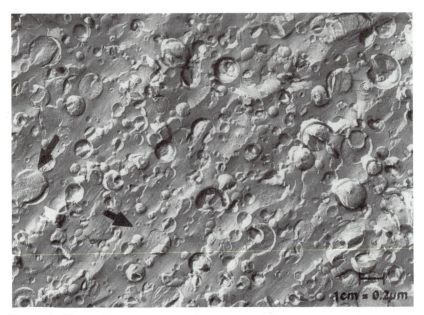

Figure 1. Freeze-fracture transmission electron micrograph of a 90% (w/v)
perflubron emulsion at a magnification of 50,000x. An emulsion droplet,
indicated by the upper left arrow, displays a grainy appearance. Water
containing liposomes are smaller and smooth in appearance.

time. The Stokes' equation for centrifugal sedimentation is used in conjunction with an assumed proportional relationship between the absorbancy and concentration. The relationship between sedimentation time under a constant field and diameter is given by:

$$d = [18 \ \eta_0 \ \ln(x_2/x_1) \ / \ (\rho - \rho_0) \ \omega^2(t) \ t \ ]^{1/2} \tag{1}$$

where d is the particle diameter, η_0 is the viscosity of the disperse phase and x_2 and x_1 are the distance between the center of rotation and the measuring plane, and the distance between the center of rotation and the sedimentation plane, respectively. The densities of the sample and the disperse phase are ρ and ρ_0 respectively. $\omega(t)$ is the rotational angular velocity and t is the time. This optical transmission method measures particle sedimentation as the change with time of the amount of transmitted light or absorbance of the particles. The relationship between the absorbance and the size and number of particles is given by:

$$\log (I_o / I_i) = P \sum_{i=1}^{n} k \, (d_i) \, N_i \, d_i^2 \tag{2}$$

where I_0 is the intensity of the light beam at the sample and I_i is the intensity of light transmitted through the sample. The optical coefficient of cell and particle is P and $k(d_i)$ is the absorption coefficient of particle d_i. N_i is the number of particles with a diameter d_i. This equation determines particle distribution measurements on the basis of cross sectional area. It is possible to convert this data to volume-, length- or number-based distributions (2).

Sedimentation Field-Flow Fractionation has been demonstrated to be a useful particle sizing technique for macromolecules, colloids, liposomes and emulsions as well as for monitoring temporal particle growth (3,4,5). This technique utilizes an applied external centrifugal field in conjunction with a laminar flow of a mobile phase. The applied centrifugal field causes the particles to accumulate towards one wall of a thin ribbon like channel placed inside the centrifuge basket. This accumulation is opposed by diffusion of the particles resulting in a sedimentation equilibrium. Larger particles with lower rates of diffusion equilibrate closest to the accumulation wall while progressively smaller particles, with higher diffusion rates, are located further from the wall and closer to the more rapidly moving areas of the laminar flow of the carrier phase. This causes smaller particles to elute first followed by increasingly larger particles found in the lower flow regions closer to the wall. Detection is accomplished via a UV-Vis optical detector (6).

The theory of field-flow fractionation as developed by Giddings et al. is very well established (7,8,9). The retention of a particle or population of similarly sized particles can be related to λ, a dimensionless parameter:

$$t_r / t_0 = 1 / [6\lambda (\coth(1/2 \ \lambda) - 2 \ \lambda)] \tag{3}$$

Since particle size determines retention time in the SdFFF, particle diameter is related to λ and the applied field strength G by the following:

$$\lambda = (6\,kT)\,/\,(\,\pi\,w\,G\,d^3\,\Delta\rho)\qquad\qquad(4)$$

where k is Boltzmann's constant, T is temperature, w is channel thickness and $\Delta\rho$ is the difference in density between the particle's density and the density of the mobile phase. The applied sedimentation field strength is given by:

$$G = r_o\,(\pi\,\omega\,/\,3)^2\qquad\qquad(5)$$

where r_0 is the radius of the centrifuge and ω is rotational speed in revolutions per minute. From these equations it is then possible to calculate particle diameter from retention time (10).

Materials and Methods

Three forms of particle size analysis were utilized to determine a particle size distribution for the test perflubron dispersion. Having established this distribution and the response of each instrument it was possible to determine the necessary photon scattering correction factor and use it in turn to calibrate the photosedimentation instruments selected to perform routine quality control measurements. The instruments selected to perform particle size analysis were a Nicomp Model 270 for photon correlation, a HORIBA CAPA-700 for photosedimentation and the S101 Colloid/Particle Fractionator manufactured by FFFractionation Inc.

The perfluorocarbon test dispersion consisted primarily of 90% (w/v) perflubron and 4% (w/v) egg yolk phospholipid (EYP) in an isotonic phosphate buffered aqueous phase. The EYP layer has a relatively larger absorptivity in the ultraviolet range than in the visible and since optical measurements were made at 249 nm and 560 nm the detector signals were normalized to eliminate this effect.

Sample preparation for the photon correlation determination required 10 µl of the test dispersion to be dispersed in 700 µl of 0.2 µm filtered deionized water and mixed well. The Nicomp 270 was operated according to manufacturer's instructions.

The photosedimentation determination was performed using the Horiba CAPA-700 instrument. The samples were prepared by dispersing 125 µl of the test dispersion in 3.0 ml of filtered deionized water and vortexing. Measurements were made by centrifuging at a fixed rotational speed of 5000 rpm with a sedimentation distance of 5 mm. The detector operated at 560 nm. The volume based particle size distribution was calculated by instrument software.

The apparatus for the SdFFF analysis (FFFractionation, Inc.) had a centrifugal radius of 14.9 cm and channel dimensions of 89.1 cm x 1.9 cm x 0.0254 cm. A mobile phase of 0.05% (w/v) sodium dodecyl sulfate and 0.01% (w/v) NaN$_3$ was pumped at a flow of 2 ml/min. The initial field strength was 1500 rpm with a stop flow period of 5 minutes and a decay constant, t_a = 7 minutes. The SdFFF used a power programmed field decay to achieve maximum resolution across the range of perflubron particles (11). A Linear UVIS 200 detector was operated at a wavelength of 560 nm for comparison with the HORIBA instrument and also at 249 nm, a wavelength with an improved signal to noise ratio.

The perflubron concentrations of a set of monosized fractions were determined by injecting 100 μl of the neat emulsion and collecting fractions across the range of the particle size distribution. Perflubron was extracted from these fractions into isooctane and determined with a Hewlett Packard 5890 Gas Chromatograph (GC) using a 30m x 0.25 μm fused silica capillary column (DB 210) and an electron capture detector. Perflubron mass concentration was calculated from an external standard curve.

Comparison of Sizing Methods. In Figure 2, the Nicomp photon correlation data of the emulsion clearly shows the bimodal nature of the distribution (i.e. the liposomal population and the emulsion droplet population). The mass median diameter of the emulsion determined by this method is 0.27 μm. The abundance of the liposomal population is overestimated by this method because of the greater effective light scattering by this population which is a result of a greater difference in refractive index between the particles and disperse phase.

The results from the photosedimentation analysis are presented in Figure 3. This histogram represents the percent frequency of the particle size distribution. The first bar of the histogram represents the liposome population. The liposomes are incorrectly sized due to the discrepancy in densities between the emulsion particles and the liposomes. It is possible to subtract this peak and determine a median particle size of 0.41μm for the cumulative distribution of the emulsion.

Figure 4 is a fractogram generated by the SdFFF instrument. The presence of the liposomes is evident as the shoulder off the void volume preceding the broad emulsion band. The mass median particle size determined by this method is 0.25 μm. The liposomes are again incorrectly sized because the particle size distribution was calculated using the density of the perflubron. It is possible to subtract this peak and calculate the median diameter of the emulsion.

The mass median particle size results determined by photon correlation and SdFFF agree satisfactorily. There is however a large discrepancy between the results of these analyses and those of the photosedimentation determination. In addition to the median particle size discrepancy, there is also a difference in the distribution of sizes detected. Since all determinations were made using the same emulsion, the difference is due to the instrumentation, and specifically the lack of a scattering correction. The perflubron emulsion, as stated previously, is composed of both large and small particles which exhibit Rayleigh and Mie light scattering. The small particles, in comparison with the larger particles, scatter light poorly. For particles smaller than the wavelength of the detector, the detector signal is related to the mass in the detector as well as the particle diameter raised to the 6th power (12). For particles larger than the wavelength of the detector light, the detector response is simply related to the particle mass concentration in the detector (13). The result of a distribution composed of such a range of particle sizes is an observed distribution biased to the larger sized particles. This problem is intensified at larger wavelengths (i.e., 560 nm vs. 249 nm). Also influencing the particle size distribution, as determined by the HORIBA, is the selected optical absorption coefficient, $k(d_i)$,

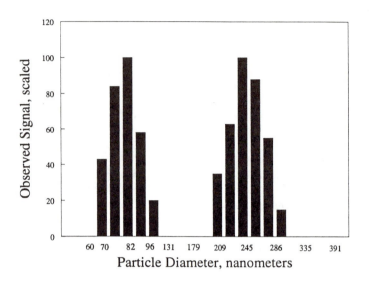

Figure 2. Volume weighted particle size distribution of a 90% (w/v) perflubron emulsion as determined by photon correlation spectroscopy. The vesicles are the peak centered at about 80 nm.

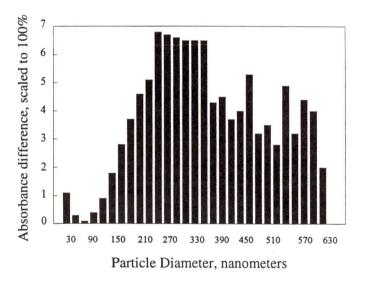

Figure 3. Percent frequency particle size distribution of a 90% (w/v) perflubron emulsion as determined by photosedimentation. The vesicles are represented by the smallest bar.

which is used to correct for the scattering effects of variously sized particles. The selection of this factor affects the calculated particle size distribution (14).

In terms of length of analysis and ease of operation, photon correlation and photosedimentation are fast and simple although photon correlation is complicated by the possible presence of dust which makes analyses more difficult. Photosedimentation is also highly reproducible with precision in the determination of median diameter being 4%. Although SdFFF analysis may be lengthy and the instrument requires more attention, this technique alone, of the three mentioned, allows for the collection of monosized fractions which can be independently and quantitatively analyzed, in this case by GC, for the determination of perflubron concentration. For this reason, SdFFF was selected as the instrument to calibrate and determine a scattering correction factor. This factor could then be used in turn to calibrate the relatively quick and reproducible photosedimentation instruments used for the routine quality control measurements.

Calibration of SdFFF. Figure 5 shows results of the fractionation experiment and subsequent GC analysis. SdFFF detector response at 249 nm and 560 nm, as well as the GC area counts are presented. The liposomal peaks have been removed and the data normalized to a fixed peak height for simplicity of comparison. This comparison shows the displacement between the mass concentration particle distribution with those that are optically determined. This figure clearly shows the difference between visible and UV detection. This shift is due to scattering. It is the scattering effects as evidenced by this data which need to be corrected for. In order to determine the appropriate correction factor for light scattering, the concentration of perflubron, as GC area, was ratioed to the SdFFF absorbance at each corresponding particle size. This factor was then multiplied by the observed absorbance at the appropriate particle size to yield a value which is directly proportional to the concentration of perflubron. These values can then be normalized and plotted to give the corrected particle size distribution. Figure 6 shows the correction factors determined at 560 and 249 nm. The factors have been scaled to approach 1.0 as the particles become larger than the wavelength of light. This is in qualitative agreement with the way particles of varying sizes scatter light of various wavelengths. Figure 7 is a representation of the raw GC normalized mass concentration plotted along with the HORIBA photosedimentation data with the correction factor applied. There is very good agreement between the two sets of data.

Conclusions

Sedimentation field-flow fractionation has been proven to be a valuable tool for measuring the particle size distribution of concentrated perflubron dispersions. The lack of an appropriate commercially available external standard necessitates calibration with the sample dispersion. Because of the unique opportunity to collect and analyze separated monosized fractions of the test dispersion by an independent analytical method, it is possible to verify the perflubron mass at discrete particle diameters. Calibration factors have been determined and are a function of detector wavelength, but beyond this are independent of instrument operating conditions.

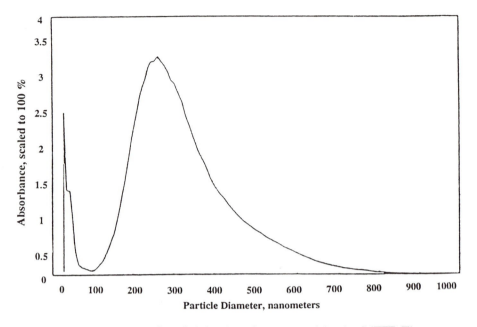

Figure 4. Fractogram of a 90% (w/v) perflubron emulsion by SdFFF. The vesicles can be seen as the shoulder off the initial sharp void volume peak. Run conditions were: field strength 1500 rpm, stop-flow time 5 min, flowrate 2.00 ml/min.

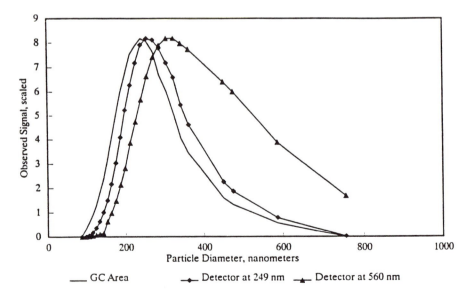

Figure 5. Comparison of mass-based particle size distribution with absorbance-based distributions at 249 nm and 560 nm. Detector signals were normalized to eliminate EYP absorption effects. Note the shift in peak maximums indicative of light scattering effects at the two wavelengths.

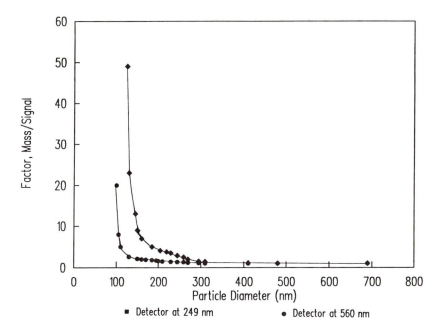

Figure 6. Absorbance-to-mass correction factors. Multiplying the background corrected absorbance by these diameter dependent factors yields a number proportional to mass.

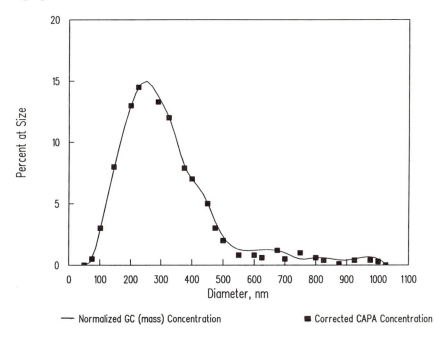

Figure 7. The normalized GC (mass) concentration with an overlay of the corrected photosedimentation particle size distribution.

With these correction factors in place, SdFFF is a versatile and powerful tool for the determination of particle size distributions in a complex sample matrix and provides calibration opportunities other techniques as well.

Acknowledgments

We wish to thank J. Zasadzinski and M. Longo of the University of California, Santa Barbara for the freeze-fracture transmission electron microscopy work.

Literature Cited

1. Nicoli, D.F., McKenzie, D.C., Wu, J.-S. *Am. Laboratory.* **1991**, Nov, 32-40.
2. Horiba Instruments Incorporated, Instruction/Service Manual.
3. Yang, F.S., Caldwell, K.D., Giddings, J.C., Astle, L. *Anal. Bio.* **1984**, *135*, 488.
4. Schallinger, L.E., Kaminski, L.A. *Bio Techniques.* **1985**, *3*, 124.
5. Kirkland, J.J., Yau, W.W., Szoka, F.C. *Science*, **1982**, *215*, 296.
6. Kirkland, J.J., Yau, W.W. *Science*, **1982**, 218, 121.
7. Giddings, J.C., Myers, M.N., Caldwell, K.D., Fisher, S.R. In *Methods of Biochemical Analysis*; Glick, D., Ed.; Wiley: New York, 1980, Vol 26; pp 79-136.
8. Giddings, J.C., *Anal. Chem.* **1981**, *53*, 1170-1175.
9. Giddings, J.C., Caldwell, K.D., Jones, H.K. In *Particle Size Distribution: Assessment and Characterization*; ACS Symposium Series 332; Provder, T., Ed.; American Chemical Society: Washington DC., 1987; 215-230.
10. Yang, F.J.F., Myers, M.N., Giddings, J.C. *Anal. Chem.* **1974**, 46, 1924-1930.
11. Williams, P.S., Giddings, J.C., *Anal. Chem.* **1987**,59, 2038
12. Van de Hulst, H.C. *Light Scattering by Small Particle;* Dover Publications, Inc., New York, 1957; 391.
13. Giddings, J.C., Moon, M. H., Williams, P. S., Myers, M.N. *Anal Chem.* **1991**, 63, 1366-1372.
14. Beckers, G.J.J., Verigna, H.J. *Powder Technology.* 1989, 60, 245-248.

RECEIVED June 5, 1992

Chapter 2

Particle-Size Analysis Using Flow Field-Flow Fractionation

S. Kim Ratanathanawongs and J. Calvin Giddings

Field-Flow Fractionation Research Center, Department of Chemistry, University of Utah, Salt Lake City, UT 84112

Flow field-flow fractionation (flow FFF) is one of the most universal separation techniques, being applicable to virtually all particles and macromolecules from a few nanometers to over 50 micrometers size. Following an introduction to flow FFF operation and principles, a comparison of flow FFF and sedimentation FFF is provided here to show the basis for the more effective applicability of flow FFF to nanometer-sized particles. Applications to seed latexes along with latex standards, fumed silica, chromatographic silica supports, and pollen grains are shown. The difficulty of simultaneously analyzing submicron and supramicron sized particles is explained and a partial remedy is proposed that involves shifting the steric transition diameter d_i up or down from its typical one μm value. By using a thin (94 μm) channel and high flowrates, it is shown that d_i can be reduced to 0.3 μm, thus expanding the range of steric mode operation.

Field-flow fractionation (FFF) is a broad class of elution methods used to separate and characterize macromolecules and particles. FFF is characterized by the use of an externally applied field acting perpendicular to the direction of a flowstream carrying components through a thin channel. The field, by driving components into different stream lamina with different velocities, induces the differential elution of the components. Different types of "fields" (e.g., gravitational, flow, electrical, etc.) and operating modes have been used, thus giving rise to different subtechniques of FFF with diverse capabilities (1-3).

In the case of flow FFF, the driving force that acts perpendicular to the axis of separation (coincident with the channel axis) is provided by a second (*crossflow*) stream of carrier entering and exiting the channel through permeable walls. This crossflow force F_f is given by Stokes law

$$F_f = 3\pi\eta d_s U \tag{1}$$

where η is the viscosity of the carrier, d_s is the Stokes diameter of the particle, and U is the transverse displacement (crossflow) velocity. Equation 1 shows that the interaction between the crossflow and entrained sample particles is determined by the effec-

0097–6156/93/0521–0013$06.00/0

tive (Stokes) diameter of the particles alone, thus producing a separation based on particle size but not on density. As a result, flow FFF has an advantage over sedimentation-based techniques (which separate based on both size and density) in the size characterization of samples whose components have varying densities or whose densities might be unknown.

Particles in the size range from about 0.002 to 50 µm can be separated and characterized by existing flow FFF systems. Figure 1 summarizes the different sizes and types of particles that have been characterized by flow FFF to date. (This figure excludes numerous macromolecular materials that have been fractionated by flow FFF including synthetic cationic and anionic water soluble polymers along with humic materials, proteins, and DNA.) Flow FFF is potentially applicable to a much broader range of inorganic, biological, and environmental colloids than shown in Figure 1.

The object of this report is to briefly review the principles of flow FFF and then to show theoretically why flow FFF is applicable to particles of much smaller size (potentially down to 1 nm or less) than sedimentation FFF. Some applications are shown that include nanometer-sized seed latex particles and, at the other end of the size spectrum, chromatographic supports. We also explain the importance of expanding the size range of steric mode operation by lowering the steric transition diameter. A strategy for achieving this goal is outlined and experimental confirmation provided.

Normal Mode. Two different modes of flow FFF operation, as described below, are used at the two ends of the particle size spectrum. Both modes can be realized using the same FFF channel. In this operating mode, applicable to the lower end of the particle size spectrum, crossflow displacement is balanced against diffusion. The displacement of particles by the crossflow of carrier liquid transports them towards the accumulation wall (semipermeable membrane) as shown in Figure 2a. The opposing diffusion gives rise to an exponential concentration profile for each component with the highest concentration at the wall. Smaller particles have a thicker profile and will extend further into the parabolic flow stream than larger particles; thus they are displaced more rapidly by channel flow and elute earlier.

The retention parameter λ is equal to the ratio of the mean thickness of the particle cloud ℓ to the channel thickness w. For flow FFF (4)

$$\lambda = \frac{\ell}{w} = \frac{kTV^0}{3\pi\eta V_c d_s w^2} \tag{2}$$

where k is the Boltzmann constant, T is the temperature, V^0 is the void volume, and $\dot{V}_c$ is the cross flowrate. The retention ratio R--the ratio of mean particle velocity to the average carrier velocity--can be expressed in terms of the void time t^0, the retention time t_r, and λ as follows (5)

$$R = \frac{t^0}{t_r} = 6\lambda\left(\coth 2\lambda - \frac{1}{2\lambda}\right) \cong 6\lambda \tag{3}$$

From equations 2 and 3, it is apparent that small particles elute prior to large particles.

Particles undergoing normal mode migration and separation will experience random Brownian displacements across their exponential concentration profile and thus across various streamlines of the parabolic flow profile. Those that diffuse towards the center of the channel will momentarily be displaced faster by the channel flow than those diffusing towards the wall. The net result of this relative displacement is the so-called *nonequilibrium band broadening*, an unavoidable (but controllable)

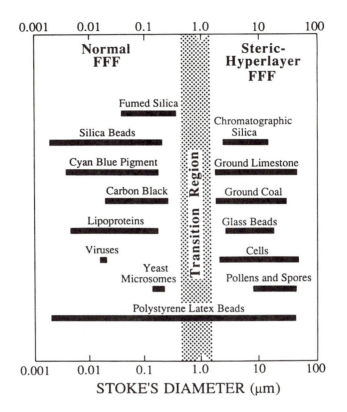

Figure 1.   Applications of flow FFF to various types of materials with different molecular weights and diameters.

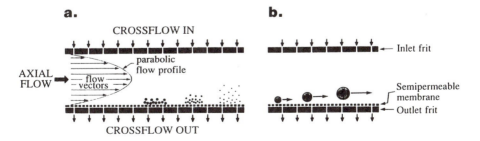

Figure 2.   Separation mechanisms in the (a) normal mode and (b) steric-hyperlayer mode.

feature of normal mode separations (6, 7). Since the nonequilibrium band spreading (measured as variance or plate height) is directly proportional to the mean fluid velocity <v>, a choice is available between low-flow conditions that minimize the band spreading and thus enhance resolution and high-flow conditions that minimize the analysis time. Band spreading is also reduced by increasing the field strength.

Normal mode flow FFF has been applied to components that span a wide range of sizes, from 300 Daltons (8) to 3 μm and above (9). The lower extreme is determined by the availability of small-pore membranes and by the high pressures that accompany the use of these membranes combined with the high cross flowrates (field strengths) needed to counteract the more vigorous Brownian motion of small species. The upper limit is determined by steric effects (see following section).

The lower size limit of flow FFF is of particular interest because this limit extends well below the diameter range for which sedimentation FFF is generally effective. Insight into the unique capability of flow FFF for separating nanometer-sized (~1-50 nm) particles can be gained from equation 2. This expression can be rearranged into the form

$$V_c = \frac{kTV^0}{3\pi\eta d_s w^2 \lambda} \tag{4}$$

which, by relating $\dot{V}_c$ to λ, gives the value of the cross flowrate $\dot{V}_c$ needed to generate a sufficiently small λ for adequate retention. According to equation 3 it is necessary that $\lambda \leq 0.1$ to drive R below approximately 0.5, the maximum R value that can be tolerated for achieving reasonable resolution. Thus taking $\lambda(max) = 0.1$, the minimum $\dot{V}_c$ from equation 4 becomes

$$\dot{V}_c(min) = \frac{10kTV^0}{3\pi\eta d_s w^2} \tag{5}$$

This equation is plotted in Figure 3 for a typical channel with $V^0 = 1.24$ mL, $w = 0.0254$ cm, and $\eta = 0.01$ poise, a value corresponding to water at 20°C. The plot, appearing on the left hand side of Figure 3, shows that the particle diameter can drop as low as 1-2 nm before $\dot{V}_c$ (min) is driven above 5 mL/min, a fairly typical and readily achievable cross flowrate used in flow FFF operation.

A parallel analysis can be carried out for sedimentation FFF. In this case the minimum field strength G needed to drive λ down to 0.1 or lower can be deduced from well known equations (10) for λ and is given by

$$G(min) = \frac{60kT}{\pi\Delta\rho w d^3} \tag{6}$$

In this case the minimum field strength $G(min)$ depends upon the difference in density $\Delta\rho$ between the particle and the carrier. Accordingly, two plots of $G(min)$ are shown in Figure 3, one for $\Delta\rho = 1.50$ g/mL and the other for 0.05 g/mL. These curves are calculated for a channel of thickness 254 μm operating at a temperature of 298 K. The rpm values shown on the right side of the plot were calculated for a sedimentation system with a rotor radius of 15.1 cm. The plots show that extremely high field strengths are needed to separate particles in the 10-50 nm size range (depending upon particle density) and demonstrate the extraordinary difficulty of resolving smaller particles even with large gains in field strength. These combined plots show that flow

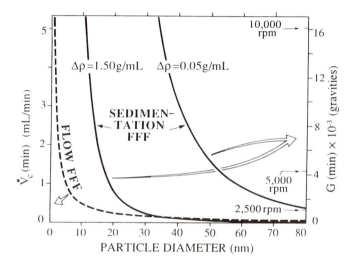

Figure 3.   Sedimentation and flow FFF forces required to retain particles of different diameters at a constant λ value of 0.1. A rotor radius of 15.1 cm was used to calculate the sedimentation FFF rpm values.

FFF is much more readily adapted to the analysis of small particles in the range 1-20 nm (and in some cases up to 50 nm) than sedimentation FFF.

Steric-Hyperlayer Mode. In the steric mode, large (>1 μm) particles are driven against the accumulation wall by the externally applied field (*11*). Their relatively small diffusion coefficients nullify significant back diffusion. Consequently the particle center ends up little more than one particle radius away from the wall. Since larger particles protrude further into the parabolic flow profile than their smaller counterparts, they occupy faster flowing laminae and elute first. Thus there is a reversal in elution order as compared to the normal mode where the smallest particles elute first.

In practice, these large particles are acted on by hydrodynamic lift forces that propel the particles away from direct contact with the wall (*11, 12*). As shown in Figure 2b, the crossflow force and the opposing lift forces drive these particles into thin focused bands (or hyperlayers) elevated a small distance above the accumulation wall. We therefore refer to this mode as the combined steric-hyperlayer mode. (Technically, the steric mode is realized when the distance of closest approach between the particle and the wall is less than the particle radius; the hyperlayer mode is considered applicable when this distance exceeds the particle radius. Flow FFF is usually operated in the hyperlayer mode.)

Hyperlayers are formed at positions where the two opposing forces are equal. Larger particles, whose centers of mass equilibrate further from the accumulation wall than those of smaller particles, emerge first in the elution sequence. The larger the two opposing forces, the smaller the particle diffusive displacements and the thinner the hyperlayer. With sufficiently thin hyperlayers, nonequilibrium band broadening is less significant than in the normal mode and fast flowrates can be used to achieve high speed separations without inordinate band spreading. This is a major advantage of steric-hyperlayer over normal mode FFF separations.

The retention time t_r in steric-hyperlayer FFF is related to particle diameter d by

$$t_r = \frac{wt^0}{3\gamma d} \tag{7}$$

where γ is a *steric correction factor* (of order unity) that compensates for lift forces and other hydrodynamic factors (*12*). The inverse relationship between t_r and d is shown by this equation. For nonspherical particles d is an effective hydrodynamic diameter that may differ slightly from the Stokes diameter d_s.

The right hand side of Figure 1 summarizes the applications of steric-hyperlayer flow FFF. The upper end of the diameter scale (100 μm) is not a firm limit and could be increased by using a thicker channel and suitable flowrates. To date, the largest particle we have retained by flow FFF is a 60 μm polystyrene latex bead using a 254 μm thick channel.

The shaded area shown in Figure 1 represents the transition region between the normal and steric-hyperlayer modes. Generally, a particle size of 1 μm is considered to demarcate normal mode from steric-hyperlayer mode separation. However, in practice the normal mode mechanism can be driven up to ~3 μm (*9*) and the steric-hyperlayer mode can be extended down to ~0.3 μm depending on the experimental conditions employed. The latter case will be discussed in more detail in a later section.

Experimental

The flow FFF system utilized here consists of a commercial channel-flow pump (Spectra-Physics Isochrom LC pump, San Jose, California), a syringe crossflow pump (built in-house), a Rheodyne 7010 injection valve (Cotati, California) with a 20 μL sample loop, a flow FFF channel, and a UV-vis detector (Spectroflow 757, Applied Biosystems, Ramsey, New Jersey) usually set at 254 nm. The flow FFF channel is assembled from two Plexiglass blocks with inset ceramic frits, a spacer from which the channel configuration has been cut and removed, and a membrane whose function is to retain components in the channel and to serve as the accumulation wall *(14, 15)*. The membranes that have been utilized in this work include ultra-filtration membranes from the Amicon (Danvers, Massachusetts) YM series and the polypropylene type microfiltration membrane designated as Celgard 2400 (Hoechst Celanese, Separations Products Division, Charlotte, North Carolina). The choice of membrane depends on the sample being characterized and the conditions used. In particular, the pore size must be small enough to retain the sample components. A variety of channels were used to test their viability in the analysis of different samples under different conditions. The various membranes and channel dimensions used in this work are summarized in Table I.

The carrier used in this study was predominantly doubly distilled deionized water containing 0.1% FL-70 (Fisher Scientific, Fair Lawn, New Jersey) and 0.02% sodium azide. However, a 0.001 M NH_4OH solution was used in the fumed silica work and an Isoton II (Coulter Diagnostics, Hialeah, Florida) solution was used for pollen separation.

Table I. Summary of Carrier, Membranes, and Channel Dimensions Used for Analysis of Different Samples

Sample	Figure no.	Carrier	Membrane	Channel[1] thickness (cm)	Void vol.[2] (mL)
PS latex beads	4	0.1% (w/v) FL-70 + 0.02% (w/v) NaN_3	Celgard 2400	0.0127	0.62
Latex seeds	5	0.1% (w/v) FL-70 + 0.02% (w/v) NaN_3	YM10	0.0230	1.12
Fumed silica	6,7	0.001 M NH_4OH	Celgard 2400	0.0127	0.62
PS latex beads	8	0.1% (w/v) FL-70 + 0.02% (w/v) NaN_3	YM30	0.0087	0.43
Chromatographic silica	9a	0.1% (w/v) FL-70 + 0.02%(w/w) NaN_3	YM30	0.0087	0.43
Pollen grains	10	Isoton II	YM30	0.0220	1.07
Latex beads	11,12	0.1% (w/v) FL-70 + 0.02% (w/v) NaN_3	YM10	0.0094	0.46

[1]For all systems, the channel breadth is 2.0 cm and tip-to-tip length is 27.2 cm.
[2]Measured void volume.

The polystyrene latex standards were obtained from Duke Scientific (Palo Alto, California) and Seradyn (Indianapolis, Indiana). The standards were diluted prior to injection to achieve suspensions that were 0.05%-0.1% solids. Cab-O-Sil and chromatographic silicas were supplied by Cabot Corp. (Tuscola, Illinois) and Phenomenex Inc. (Torrance, California), respectively. The sample concentrations were 25 mg/mL for Cab-O-Sil and 5 mg/mL for chromatographic silicas. The ragweed and paper mulberry pollens were obtained from Duke Scientific and the pecan pollen from Polysciences (Warrington, Pennsylvania). The pollen sample used in this work was comprised of 1-2 mg/mL of each pollen.

Results and Discussion

A number of applications of flow FFF to particle separation and characterization are described below. Both normal and steric-hyperlayer mode applications are shown involving primarily submicron-sized and supramicron-sized particles, respectively. However in this presentation we emphasize by means of several examples and background theory the capability of flow FFF to work effectively on particle populations whose sizes extend toward the lower limits of the respective ranges. Thus for normal mode FFF, we demonstrate the applicability of this approach to nanometer size particles whose distributions are very difficult to characterize by other means. With regard to the steric-hyperlayer mode of flow FFF, we show that the lower limit can be pushed down to diameters of 0.3-0.4 μm, which opens up the possibility of analyzing particle populations extending from this lower limit up to 10 μm, or perhaps 20 μm, without concern for retention inversion. These developments will be explained in more detail below.

Normal mode flow FFF. To illustrate the capability of flow FFF to resolve polymer latexes and related particles, polystyrene latex standards having narrow size distributions were used as samples. A typical flow FFF separation is shown in Figure 4. The flow FFF channel used for this run comprised a 127 μm thick spacer and a polypropylene membrane (see Table I). The channel flowrate $\dot{V}$ was 4.90 mL/min and the cross flowrate $\dot{V}_c$ was 2.65 mL/min. A stopflow time of 0.25 min was used. Baseline resolution was obtained for the three components in just over 20 minutes. Similar latex separations were demonstrated in a previous publication (*16*).

Flow FFF in the normal mode is especially suited for the size analyses of nanometer-sized (~1-50 nm) particles. This suitability derives from the ease with which the cross flowrate is adjusted to particle diameter in accordance with the requirements of equation 5. The cross flowrates needed to retain nanometer-sized particles such that $\lambda = 0.1$ are easily applied--for example, $\dot{V}_c$ (min) = 0.47 mL/min for a 10 nm particle (see Figure 3). It is evident from equation 5 and Figure 3 that $\dot{V}_c$ can be further increased to retain even smaller sample components, including synthetic and biological macromolecules (*17, 18*). The density-independence of flow FFF retention means that the same cross flowrate requirement applies to particles of different densities, including particles that are neutrally buoyant in the carrier liquid.

Figure 5 shows fractograms and normalized particle size distribution curves for two different types of seed latex particles--polystyrene and styrene-acrylic. The flow FFF channel assembly used in this work consisted of an Amicon YM10 ultrafiltration membrane and a 254 μm thick spacer. (Void volume measurements indicated that the actual channel thickness was 230 μm; the observed difference in spacer and channel thickness is due to the compressibility of the ultrafiltration membrane.) The flowrates were $\dot{V} = 1.88$ mL/min and $\dot{V}_c = 1.00$ mL/min. A lower channel flowrate would have produced less nonequilibrium band broadening but also longer analysis times. As shown in the fractograms of Figure 5a, the analysis times are of the order of 10 minutes (excluding the 1.5 min stopflow time). One advantage of normal mode FFF

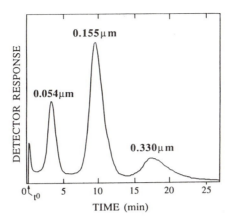

Figure 4.   Normal mode separation of submicron polystyrene latex standards.

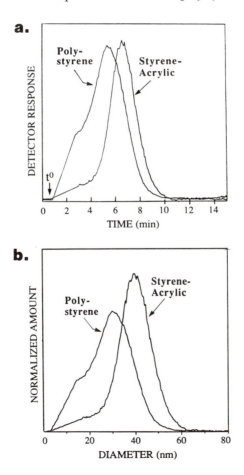

Figure 5.   Normal mode analysis of latex seed particles. (a) Fractograms and (b) particle size distributions.

is the existence of explicit theory relating retention times to diameters; there is no need for assumptions about particle density as required in sedimentation FFF nor for calibration standards (*1*). Using this established theory, the fractograms of Figure 5a are readily converted to the size distributions shown in Figure 5b.

Flow FFF can also be used to characterize populations of nonspherical particles such as fumed silica (*19*). (Fumed silica consists of branch-like structures formed from the aggregation of silica spheres.) Figure 6 shows the fractograms and corresponding particle size distribution curves of three different types of Cab-O-Sil fumed silica: EH-5, M-5, and L-90. The flowrates employed were $\dot{V} = 0.50$ mL/min and $\dot{V}_c = 0.41$ mL/min and the stopflow time was 1.7 min. The channel thickness in this case was 127 μm and the membrane was Celgard 2400. The diameters shown in the size distribution curves are, of course, Stokes diameters.

The analysis of the fumed silica sample L-90 was carried one step further. Fractions were collected in the course of an L-90 run at the time intervals indicated in Figure 7. Each fraction was then examined by scanning electron microscopy. The micrographs show the structural nature of the fumed silica in each fraction. The coupling of flow FFF and SEM has the unique capability of correlating particle structure with its hydrodynamic size. Since fractions can be readily collected in the course of FFF runs, flow FFF can similarly be coupled with other tools (e.g., ICP-mass spectrometry) to correlate a variety of properties (e.g., elemental composition) with particle size.

Steric-Hyperlayer Mode. An example demonstrating large particle fractionation by steric-hyperlayer FFF is shown in Figure 8. Monodisperse polystyrene standards (20, 15, 10, 7, and 5 μm) were eluted as narrow peaks within six minutes. The channel consisted of a 127 μm thick spacer and a YM30 membrane. The flowrates used were $\dot{V} = 1.36$ mL/min and $\dot{V}_c = 2.49$ mL/min and the stopflow time was 0.3 min. Analysis times can be significantly reduced by using higher channel flowrates since nonequilibrium band broadening is relatively small. Thus separation in the steric-hyperlayer mode can be very fast, for example, 6 sec for the partial separation of 49, 30, and 20 μm latex beads (*15*).

Unlike normal mode FFF, calibration curves are necessary in the steric-hyperlayer mode because lift forces and their effects on retention times have not been fully characterized. In the calibration procedure, a run such as that shown in Figure 8 is made using standards that approximately bracket the size range of the sample. When the logarithm of the retention time of different size standards is plotted against the logarithm of their respective diameters, a straight line is usually obtained. Using this calibration plot, the diameters of unknown samples or fractions can be found and size distributions obtained (*14*).

Chromatographic supports have been analyzed by this method (*14*). Retention depends on particle size alone and is independent of porosity, unlike results from sedimentation FFF or electrozone methods. To illustrate this concept, the fractograms obtained for two new supports, Optisil 10 SCX and Selectosil 5 SCX, are shown in Figure 9a. The run conditions are identical to those used for the calibration run shown in Figure 8. The diameter corresponding to any observed retention time can be obtained from the appropriate calibration plot such as that shown in Figure 9b based on the fractogram of Figure 8. The times corresponding to the first moments of the eluted chromatographic support peaks were used here. From these values of t_r, the corres-ponding diameters d could be obtained from the straight line calibration plot of Figure 9b. The FFF measured diameters of 12 and 6 μm indicate that these silica particles are larger than their nominal values of 10 and 5 μm for Optisil and Selectosil, respectively. The percent coefficient of variation (equal to the standard deviation divided by the mean particle diameter × 100) calculated directly from the peak widths of the fracto-grams (*14*) are found to have values of 25% for Optisil and 23% for Selectosil.

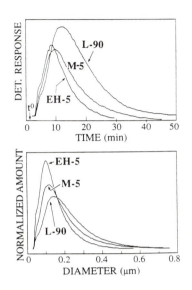

Figure 6.   Fractograms (top) and particle size distributions (bottom) of different types of Cab-O-Sil fumed silica.

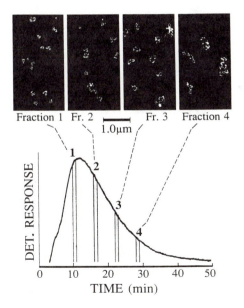

Figure 7.   Scanning electron micrographs of Cab-O-Sil L-90 fractions collected at the indicated time intervals.

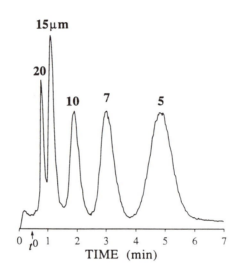

Figure 8.   Steric-hyperlayer mode separation of polystyrene standards.

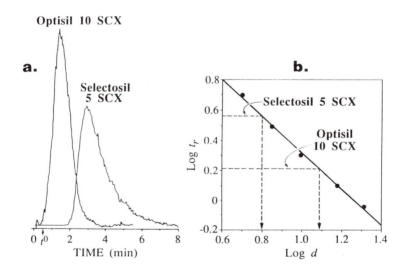

Figure 9.   (a) Fractograms of the two silica based chromatographic supports Optisil 10 SCX and Selectosil 5 SCX and (b) acquisition of log d values from calibration curve and from retention times of peak centers of gravity shown in part (a).

Another example is the rapid separation of different types of pollen grains as shown in Figure 10. The channel employed was a frit-inlet flow FFF channel (*20*) with dimensions of 38.2 cm × 2 cm × 0.022 cm. (Frit-inlet hydrodynamic relaxation was introduced as an alternate approach to the stopflow procedure to induce the rapid relaxation of particles to their equilibrium positions (*20*).) In this method, a frit-inlet flowrate $\dot{V}_f$ introduced through a frit element of the depletion wall near the inlet drives incoming sample components rapidly toward their equilibrium positions near the accumulation wall without the need to stop the axial flow. The flowrates were $\dot{V}_f$ = 3.6 mL/min, sample introduction flowrate = 0.4 mL/min, and $\dot{V}_c$ = 6.8 mL/min. The optical micrographs shown in Figure 10 confirm the separation of the different types of pollen grains. The smooth spheres are polystyrene standards (47.8 µm, 29.4 µm, and 19.6 µm) that had been mixed in with the pollen grains to demonstrate the density independence of flow FFF.

Extending the Steric-Hyperlayer Range. In the normal mode of FFF, retention time t_r increases with particle diameter as described by equations 2 and 3. In the steric and hyperlayer modes, retention time increases with decreasing particle diameter, as shown by equation 7. At some critical diameter these opposing trends converge and the transition between modes occurs. It is difficult to deal with populations of particles that span across this inversion region because in that case a single retention time might correspond to two different particle diameters. The relative amounts of the two intermingled subpopulations cannot then be readily discerned.

The steric-hyperlayer inversion point (the point of transition between the normal mode and the steric-hyperlayer mode) generally occurs in the vicinity of 1 µm. However, the inversion diameter d_i is dependent on experimental conditions (*21, 22*). Thus d_i can in principle be altered to fit experimental needs. When dealing with particulate samples whose diameters span across the normal inversion region (around 1 µm), some of the particles will elute in the normal mode and others in the steric mode, thus intermingling the two subpopulations. In this case d_i can be shifted up or down to approach one of the extremities of the size distribution so that most particles will elute under the governance of a single operating mode, thus reducing or eliminating the intermingling of different sizes of particles. It is generally more promising to lower d_i to expand the coverage of the steric-hyperlayer mode than to raise d_i and expand the range of normal mode fractionation. (The latter has been shown to be possible with the normal mode extending up to and beyond 3 µm (*9*).) Steric-hyperlayer operation reduces the nonequilibrium effect which is the major source of intrinsic band broadening in the normal mode. Since nonequilibrium band broadening increases with increasing flow velocity (*6, 7*), a practical limit is placed on the maximum channel flowrate that can be employed in the presence of nonequilibrium effects. This limit is higher for the steric mode than for the normal mode, thus allowing faster operation in the steric mode (see Figures 8-10).

One approach to expanding the size range of the steric-hyperlayer mode is through field and flow programming, as recently shown (*23*). Another is described below.

The hydrodynamic lift forces whose effects are observed in steric-hyperlayer experiments can be used to elevate submicron particles into hyperlayers above the accumulation wall. As a result, these submicron particles, ordinarily eluting in the normal mode, are driven into the hyperlayer mode. Thus d_i is reduced. (The larger the lift force, the smaller the particle that would undertake hyperlayer migration, and the smaller the resultant d_i.) The magnitude of the lift forces is known to increase with shear rate (*12*). In order for the effects of lift forces to be observed on increasingly smaller particles, the shear rate must therefore be increased.

The shear rate at the wall of an FFF channel can be expressed by (*12*)

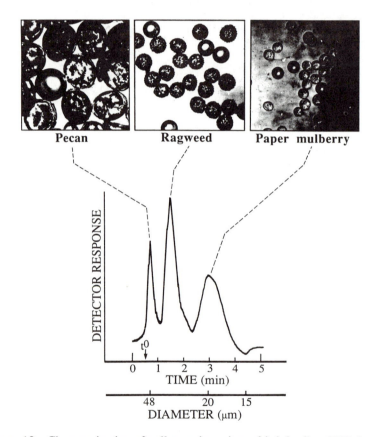

Figure 10. Characterization of pollen grains using a frit-inlet flow FFF channel.

$$s_0 = \frac{6<v>}{w} = \frac{6V}{bw^2} \tag{8}$$

where $<v>$ is the mean carrier flow velocity, $\dot{V}$ is the flowrate, and b is the channel breadth. Clearly to increase the magnitude of the lift forces through their shear rate dependence, one must increase $\dot{V}$ or diminish b or w. Since s_0 is most sensitive to w, we have elected to use a channel of reduced thickness to examine the possibility of significantly reducing d_i.

A thin channel was assembled consisting of a 127 μm thick spacer and an Amicon YM10 membrane. Void volume measurements showed the actual channel thickness to be 94 μm. The flowrate conditions were set at $\dot{V}$ = 4.22 mL/min and $\dot{V}_c$ = 1.65 mL/min. The steric mode separation of 2.062, 1.05, and 0.494 μm polystyrene latex beads was accomplished in 8 minutes as demonstrated in Figure 11. Under these same conditions, it is possible to observe the transition from the normal mode to the steric-hyperlayer mode by injecting standards of various diameters and plotting the measured retention ratio R versus the particle diameter. Such a plot is shown in Figure 12 (with particle diameters of 3.009, 2.062, 1.05, 0.868, 0.742, 0.596, 0.494, 0.426, 0.330, 0.272, 0.232, and 0.198 μm). The normal mode region is characterized by the negative slope of the R versus d plot, signifying the elution of small particles prior to large. The steric region is represented by a positive slope, indicating a reversal in elution order in which large particles elute earlier than small particles. The minimum found between these two regions corresponds to the inversion point. The inversion diameter d_i was observed to have been reduced to 0.33 μm.

The transition region can likely be further shifted by changing experimental conditions and channel dimensions more severely along the lines discussed above. This would be of considerable value in the analysis of samples whose particles extend both below and above 1 μm diameter.

Legend of Symbols

b	channel breadth	T	temperature
d	particle diameter	t^0	void time
d_i	inversion diameter	t_r	retention time
d_s	Stokes diameter	U	transverse displacement velocity
F	force	$<v>$	mean fluid velocity
F_f	crossflow force	$\dot{V}$	channel flowrate
G	sedimentation field strength	V^0	void volume
k	Boltzmann constant	$\dot{V}_c$	cross flowrate
ℓ	thickness of particle cloud	$\dot{V}_f$	frit-inlet flowrate
R	retention ratio	w	channel thickness
s_0	shear rate at wall		

Greek

γ	nonideality factor
$\Delta\rho$	difference between particle and carrier density
η	carrier viscosity
λ	retention parameter

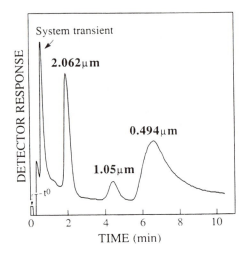

Figure 11. Fractogram of polystyrene latex beads demonstrating the extension of the steric-hyperlayer mode into the submicron size range.

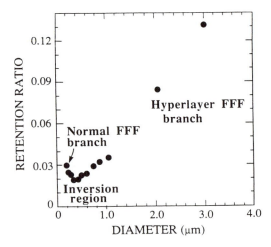

Figure 12. Plot of retention ratio versus diameter illustrating the steric-hyperlayer transition.

Acknowledgments

This work was supported by Grant CHE-9102321 from the National Science Foundation.

Literature Cited

1. Giddings, J. C. *Chem. Eng. News* **1988**, *66* (October 10), 34-45.
2. Caldwell, K. D. *Anal. Chem.* **1988**, *60*, 959A-971A.
3. Beckett, R. *Environ. Tech. Lett.* **1987**, *8*, 339-354.
4. Giddings, J. C.; Lin, G. C.; Myers, M. N. *J. Colloid Interface Sci.* **1978**, *65*, 67-78.
5. Grushka, E.; Caldwell, K. D.; Myers, M. N.; Giddings, J. C. *Sep. Purif. Methods* **1973**, *2*, 127-151.
6. Giddings, J. C.; Yoon, Y. H.; Caldwell, K. D.; Myers, M. N.; Hovingh, M. E. *Sep. Sci.* **1975**, *10*, 447-460.
7. Smith, L. K.; Myers, M. N.; Giddings, J. C. *Anal. Chem.* **1977**, *49*, 1750-1756.
8. Beckett, R.; Bigelow, J. C.; Zhang, J.; Giddings, J. C. In *Influence of Aquatic Humic Substances on Fate and Treatment of Pollutants*; MacCarthy, P.; Suffet, I. H., Eds.; ACS Advances in Chemistry Series No. 219; American Chemical Society: Washington, D.C., 1988; Chapter 5.
9. Jiang, Y. Field-Flow Fractionation Research Center, University of Utah, unpublished results.
10. Giddings, J. C.; Karaiskakis, G.; Caldwell, K. D.; Myers, M. N. *J. Colloid Interface Sci.* **1983**, *92*, 66-80.
11. Giddings, J. C.; Myers, M. N. *Sep. Sci. Technol.* **1978**, *13*, 637-645.
12. Williams, P. S.; Koch, T.; Giddings, J. C. *Chem. Eng. Commun.* **1992**, 111, 121-147.
13. Barman, B. N.; Myers, M. N.; Giddings, J. C. *Powder Technol.* **1989**, *59*, 53-63.
14. Ratanathanawongs, S. K.; Giddings, J. C. *J. Chromatogr.* **1989**, *467*, 341-356.
15. Giddings, J. C.; Chen, X.; Wahlund, K.-G.; Myers, M. N. *Anal. Chem.* **1987**, *59*, 1957-1962.
16. Ratanathanawongs, S. K.; Giddings, J. C. In *Particle Size Distribution II: Assessment and Characterization*; Provder, T., Ed.; ACS Symposium Series No. 472; American Chemical Society: Washington, DC, 1991; pp 229-246.
17. Wahlund, K.-G.; Winegarner, H. S.; Caldwell, K. D.; Giddings, J. C. *Anal. Chem.* **1986**, *58*, 573-578.
18. Giddings, J. C.; Benincasa, M. A.; Liu, M.-K.; Li, P. *J. Liq. Chromatogr.*, in press.
19. Giddings, J. C.; Ratanathanawongs, S. K.; Barman, B. N.; Moon, M. H.; Liu, G.; Tjelta, B. L.; Hansen, M. E. In *Colloid Chemistry of Silica*; Bergna, H., Ed.; ACS Advances in Chemistry Series No. 234; American Chemical Society: Washington, DC, accepted.
20. Liu, M.-K.; Williams, P. S.; Myers, M. N.; Giddings, J. C. *Anal. Chem.* **1991**, *63*, 2115-2122.
21. Myers, M. N.; Giddings, J. C. *Anal. Chem.* **1982**, *54*, 2284-2289.
22. Lee, S.; Giddings, J. C. *Anal. Chem.* **1988**, *60*, 2328-2333.
23. Ratanathanawongs, S. K.; Giddings, J. C. *Anal. Chem.* **1992**, *64*, 6-15.

RECEIVED July 6, 1992

Chapter 3

Separation and Characterization of Polymeric Latex Beads and Aggregates by Sedimentation Field-Flow Fractionation

Bhajendra N. Barman[1,3] and J. Calvin Giddings[2]

[1]FFFractionation, Inc., P.O. Box 58718, Salt Lake City, UT 84158
[2]Field-Flow Fractionation Research Center, Department of Chemistry, University of Utah, Salt Lake City, UT 84112

The high resolution separation and characterization of latex beads and their aggregates by sedimentation field-flow fractionation (SdFFF) are described here. Emphasis is placed on the principles of SdFFF, power programming and associated fractionating power, optimization strategies, determination of particle size distribution, and applications. Experimental results are provided on both narrow and broad latex distributions and on latexes having densities lower as well as higher than that of the aqueous carrier liquid. The separation of polystyrene latex particles whose diameters differ only 10% verifies the extraordinary resolving power of SdFFF. Aggregates of a 0.230 μm polymethylmethacrylate latex were resolved into eight cluster sizes (from singlets to octets) using SdFFF. Scanning electron microscopy was used as a complementary technique on collected fractions to verify elution sequence and particle size.

Since initial developments in the early 1970s (1), sedimentation field-flow fractionation (SdFFF) has been applied extensively to the separation and characterization of numerous particulate samples including latex beads falling in both submicron and micron size ranges (2-5). *Normal mode SdFFF* is applicable to particles (solid or liquid) of submicron size, whereas *steric* and *hyperlayer mode SdFFF* provide the separation and characterization of particles larger than 1 μm (5-7). In this paper we will limit our discussion to the separation and characterization of submicron sized latex beads and their aggregates by normal mode SdFFF.

Polymeric latex beads are produced for various purposes. At one extreme, uniform latex beads (primarily polystyrene, PS) are used as calibration standards for various instruments such as electron microscopes and particle counters. These beads are also used as media for numerous diagnostic tests (8). By contrast, industrial applications often utilize copolymeric latexes having mixed chemical composition and broad particle size distributions. These latex dispersions are used as ingredients for commercial paints, coatings, and adhesives (9). Some of these latexes are also used in rubber and related industries (10, 11).

---

[3]Current address: Texaco Inc., P.O. Box 1608, Port Arthur, TX 77641

Many methods are applied for particle sizing and for the monitoring of particle aggregation including microscopy, light scattering, capillary hydrodynamic fractionation, X-ray and neutron scattering, and sedimentation (*12-14*). The mean particle diameter and broadness of the size distribution obtained by any method is important for controlling the consistency and quality of latex products. The characterization of latex aggregates as well as the aggregation process itself has also received considerable attention due to the fact that agglomerated particles, in general, have significant (often adverse) effects on the performance and quality of the end products. SdFFF has a number of advantages for such latex analyses including high resolution in acquiring size distribution curves, simplicity and flexibility in operation, ready automation, adaptability to different samples and different analytical needs, a clear and direct theoretical foundation, and a capability to provide narrow fractions for further characterization.

Particle analysis by normal mode SdFFF is based on the high resolution separation of particles in a thin open channel. In such a simple geometry the theoretical principles governing the separation process can be elucidated in considerable detail. This simplifies optimization. The operation of SdFFF is extremely flexible; variations in the field strength and flowrate, along with different field programming options, can be used to meet speed and resolution criteria for most submicron particle populations. The experimental procedure is simple and straightforward, resembling that utilized for chromatography.

Sample separation in SdFFF is realized in a three-step process: injection, relaxation, and elution. These steps are shown in Figure 1. With regard to step 1, the sample is injected and carried to the inlet end of the channel. Then in step 2 the flow is stopped and the driving force (if not already in place) is applied so that particles are centrifuged toward one wall. Since the particle accumulation is opposed by Brownian motion, relaxation to a steady state distribution near the wall occurs during this relaxation step. Since smaller particles have more active Brownian motion and are subjected to weaker forces, their equilibrium distribution will have a greater mean thickness than that of larger particle populations.

Following relaxation, the flow is started and sample displacement begins during the elution (third) step. Since the laminar flow between the channel walls is parabolic in form, smaller particles with an elevated mean layer thickness are carried downstream faster than those near the wall. Therefore, in the normal mode of SdFFF, the smallest particles elute first and are followed by those with increasing particle diameters. As each particle fraction elutes, it is carried into a detector which measures the quantity of particulate matter corresponding to each particular elution time.

In SdFFF, the particle elution time is determined by two important properties: particle mass (or diameter), and density. (Particle shape plays no role except in extreme circumstances.) Since SdFFF is a mass-based separation method, particles eluting at any specified retention time have a specific mass or diameter (or effective spherical diameter for irregularly shaped particles). It is shown later that SdFFF equations can be used to convert the detector response versus retention time profile to a concentration versus particle diameter curve representing the particle size distribution (PSD).

The unique capabilities of SdFFF in studying particle aggregation by means of the separation and characterization of individual clusters of different mass was first demonstrated in the separation of viral rods (*15*) and more recently in the separation of latex aggregates (*16-19*). As the separation process is described by well defined theoretical principles, it is possible to establish resolution criteria (*16*) and to determine the mass polydispersity of latex aggregates (*17*). Possible secondary effects arising from the cluster size and conformation can also be ascertained by comparing experimental results with those predicted by theory (*20*).

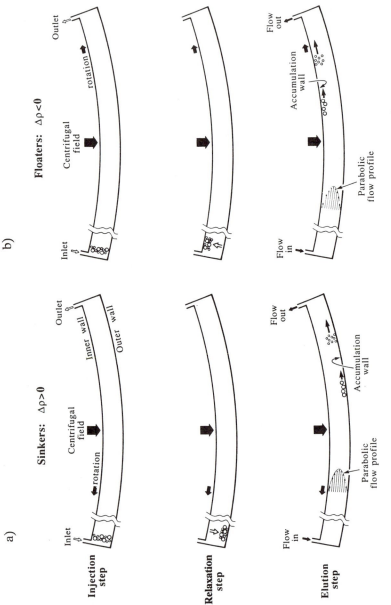

Figure 1. A schematic diagram of an edge-on view of a SdFFF channel showing particle positions during injection, relaxation, and elution steps for (a) particles denser than the carrier liquid (sinkers: Δρ > 0) and (b) particles less dense than the carrier liquid (floaters: Δρ < 0).

A number of features related to the analysis of latex beads and their aggregates by SdFFF are addressed here. In particular, the underlying principles of SdFFF and the scope and limitations of this method for the characterization of latexes having both uniform and broad particle size distributions, as well as having aggregates composed of narrow and broad latex bead populations, are summarized in this study. Experimental results are also shown for the analysis of samples having densities both lower and higher than that of water.

Theory

The theory describing the normal mode SdFFF characterization and resolution of colloids and their aggregates has been developed elsewhere (*16, 17*). However, the following points are relevant in providing a background for this work.

Particle Retention in SdFFF. The standard retention equation in field-flow fractionation relates the experimental retention volume V_r (or retention time t_r) to the channel void volume V^0 (or void time t^0) and the retention parameter λ

$$\frac{V_r}{V^0} = \frac{t_r}{t^0} = \frac{1}{6\lambda[\coth(1/2\lambda) - 2\lambda]} \tag{1}$$

For well retained particles ($V_r > 2V^0$), λ is small and the following approximations are valid

$$V_r \cong \frac{V^0}{6\lambda} \, , \, t_r \cong \frac{t^0}{6\lambda} \tag{2}$$

The parameter λ in SdFFF is related to particle mass m or effective spherical diameter d by

$$\lambda = \frac{kT}{wG(|\Delta\rho|/\rho_p)m} = \frac{6kT}{\pi wG|\Delta\rho|d^3} \tag{3}$$

where k is the Boltzmann constant, T is the absolute temperature, G is the centrifugal acceleration, w is the channel thickness, ρ_p is the particle density, and $\Delta\rho$ is the difference in density between the particle and carrier liquid.

Equations 2 and 3 can be combined to obtain

$$V_r \cong \frac{V^0 wG(|\Delta\rho|/\rho_p)}{6kT}m = \frac{\pi V^0 wG|\Delta\rho|}{36kT}d^3 \tag{4}$$

The density difference $\Delta\rho$ can be negative or positive (without affecting the value of $|\Delta\rho|$) depending on whether the particle density is lower or higher than the density of the carrier liquid. Depending on sample density, particles may accumulate at either the outer or inner channel wall during the relaxation and elution steps (see Figure 1). Floating particles or "floaters," having a density less than that of the carrier liquid will accumulate at the inner wall as shown in Figure 1. "Sinkers," particles denser than the carrier liquid, will accumulate at the outer wall. According to equation 4 neutrally buoyant particles with $\Delta\rho \cong 0$ will not be retained. Therefore the difference between particle density and carrier density must be large enough for adequate retention. In most cases, a dilute aqueous carrier with added surfactant has a density sufficiently different from that of suspended particles to provide the required density difference. For particles closer to neutral buoyancy, it is necessary to use density modifiers such as sucrose (21, 22), methanol (23), and glycerine (24) to fulfill the requirements for the density difference. It is clear from equation 4 that particles of larger size do not require as large a density difference to provide significant retention as do particles of smaller size. In practice, a density difference of about 0.03 g/cm^3 is adequate for the retention of 0.1 µm particles at 1000 gravities; smaller particles require either an enhancement in $\Delta\rho$ through density modification or an increase in field strength.

The center term of equation 4 indicates that retention volume is approximately proportional to particle mass. Since aggregates that form from uniform latex populations differ from one another by only one elementary particle mass, SdFFF should provide a repetitive series of peaks with nearly equal spacing. These peaks correspond to singlets, doublets, triplets, and so on. The effective spherical diameter of a latex cluster can be obtained by using the final term of equation 4 or more accurately by using equations 1 and 3. We note that the effective spherical diameter d_n of a cluster of n elementary particles of diameter d_1 is $d_n = n^{1/3}d_1$.

Resolution and Fractionating Power. The *resolution* R_s between sub-populations of particles in two adjacent component peaks can be defined by

$$R_s = \frac{\Delta z}{4\overline{\sigma}} \tag{5}$$

where Δz is the distance between peak centers and $\overline{\sigma}$ the mean standard deviation of the peaks from the FFF channel. Resolution has been found by theoretical analysis to depend on particle polydispersity and on nonequilibrium effects arising from channel flowrate (25). Specifically, resolution is given by

$$R_s = \frac{1}{4}\left(\frac{L}{H_p + C<v>}\right)^{1/2} \frac{\Delta m}{m} \tag{6}$$

where L is the channel length, H_p is the plate height contribution due to sample polydispersity, $<v>$ is the mean linear flow velocity, C is a constant independent of $<v>$ (representing nonequilibrium band broadening), Δm is the mean mass difference between the particles in the two neighboring peaks, and m is the mean particle mass. The polydispersity contribution to plate height H_p is related to the standard deviation in particle diameter σ_d by

$$H_p = 9LS_m\left(\frac{\sigma_d}{d}\right) \tag{7}$$

where S_m is the mass selectivity; $S_m \cong 1$ for highly retained particles in SdFFF. Note that a large σ_d (associated with a polydisperse population) significantly hinders resolution. The resolution of aggregated latex clusters is reduced because of their polydispersity. If sample polydispersity is negligible, R_s varies inversely as $<v>^{1/2}$.

If the particle size distribution is continuous rather than composed of discrete sized subpopulations, then "resolution" loses meaning because there are no unique subpopulations to resolve. Instead, to obtain accurate and detailed size distributions, every infinitesimal slice of the size distribution must be maximally resolved from all other slices. Thus a more global resolution criterion is required than that defined (see equation 5) for specific pairs of component subpopulations. A more global measurement of resolutions is provided by the *fractionating power*, described below.

We note from equation 6 that resolution R_s increases in proportion to the relative difference $\Delta m/m$ in the masses of any two components being separated. Thus the ratio of R_s to $\Delta m/m$ is approximately constant independent of Δm; its value governs the resolution of any and all pairs of components having specific $\Delta m/m$ values. This ratio, providing a global criterion of resolution, is defined as the *mass-based fractionating power* (26)

$$F_m = \frac{R_s}{\Delta m/m} \tag{8}$$

Note that F_m is a dimensionless parameter. A given value of F_m can be interpreted as providing unit resolution to subpopulations as close in relative mass as $\Delta m/m = 1/F_m$, or simply $m/\Delta m = F_m$. Thus if $F_m = 5$, components for which $m/\Delta m = 5$ (or $\Delta m/m = 0.2$, corresponding to a 20% mass difference) have unit resolution.

In order to evaluate F_m for a constant field run, equation 6 (with $H_p = 0$) is substituted into equation 8, giving

$$F_m = \frac{1}{4}\left(\frac{L}{C\langle v\rangle}\right)^{1/2} \tag{9}$$

Values of C and their dependence on field strength, diffusion coefficients, and other parameters, are available from theory (25, 27). Equation 9 does not apply to field-programmed FFF (see below), which requires different equations for F_m (26, 28).

Programmed SdFFF. According to equation 4, constant field operation (using constant G) is preferable for aggregated latex populations to obtain regularly spaced cluster peaks. However, for populations of particles covering a much broader diameter and mass range, *field programming* (where field strength G is decreased with time) is essential to minimize run time. A SdFFF analysis can be carried out under conditions of *power programming*, a unique form of field programming capable of providing uniform fractionating power according to the definition of equation 8 (28). Here the field strength G is held constant at an initial level G_0 for a time-lag period t_1 and decreased according to a power function of elapsed time t. For SdFFF, the preferred form of the power decay function is

$$G = G_0\left(\frac{t_1 - t_a}{t - t_a}\right)^8 \tag{10}$$

where t_a is a constant. A uniform fractionating power throughout the diameter range can be realized if $t_a = -8t_1$ (28). In this case the constant fractionating power is approximated by the asymptotic equation

$$F_m = 0.3116\left(\frac{t^0}{\eta}\right)^{1/2}\left[\frac{(kT)^2}{w^5}G_0\Delta\rho\left(\frac{t_1}{t^0}\right)^8\right]^{1/6} \tag{11}$$

This expression has been obtained from equation 45 of ref. 28 assuming the application of SdFFF to spherical particles. We also used the relationship

$$F_m = \frac{F_d}{3} \tag{12}$$

where F_d is the diameter-based (rather than mass-based) fractionating power. F_d is defined in the same way as F_m (see equation 8) except that $\Delta m/m$ is replaced by the relative difference in particle diameters, $\Delta d/d$. It is most appropriate to use F_d when the primary emphasis is on particle diameter and F_m when the focus is on particle mass.

We note that forms of field programming other than power programming can also be applied, including linear, parabolic, and exponential programming (*29-31*). However, only the power programming described above provides a uniform fractionating power over a broad diameter range (*32*).

Secondary Steric Effects and Steric Correction in SdFFF. It was mentioned earlier that a steady state particle layer is formed when particle Brownian motion away from the accumulation wall is balanced against the centrifugal field. The particle diameter d is usually small compared to the average layer thickness ℓ, specifically $d \ll \ell$, where ℓ is typically 2-20 μm. The layer thickness is smaller for the larger particles that are more strongly driven toward the wall. Thus the condition $d \ll \ell$ gradually loses its validity with increasing d.

We observe that the standard retention equations are based on the assumption that $d \ll \ell$. However, steric perturbations (due to the physical size of the particles) becomes important with increasing particle size or cluster size as the increasing d approaches the decreasing ℓ. Therefore, steric effects become increasingly significant for larger particle diameters and/or for higher order latex aggregates. They are obviously most important for clusters that have relatively extended configurations (*18, 20*).

The standard SdFFF equations (equations 1 and 2) can be corrected for steric effects as follows

$$V_r = \frac{V^0}{6\lambda + 3\gamma d'/w} \tag{13}$$

where d' is the "effective steric diameter" of the particles and γ is the steric correction factor of order unity (*33*). Usually, for an irregular-shaped particle or particle cluster, d' corresponds to the longest dimension of the particle.

Equation 3 and equation 13 can be combined to yield

$$V_r = \frac{AmG}{1 + B\gamma d'mG} \tag{14}$$

where A and B are constants. For relatively high field conditions and/or large particle (or cluster) size, the second term in the denominator of equation 14, reflecting steric effects, becomes significant. Under these conditions, deviation from the proportionality of retention volume to the particle mass at constant field strength (predicted by equation 4) is apparent (*16*). Similarly, retention volume is no longer expected to be proportional to field strength G for a particular cluster size of a fixed mass (*18*).

Experimental

Four different SdFFF systems, identified as I, II, III, and IV, were used in this work. Systems I, II, and III are model S101 colloid/particle fractionators from FFFractionation, Inc. (Salt lake City, UT). The channel in each system was cut out of a Mylar spacer and was sandwiched between two Hastelloy C rings with highly polished surfaces constituting the walls. The three systems varied only in the channel dimensions. System I was equipped with a channel having a void volume $V^0 = 4.47$ mL, a tip-to-tip length $L = 90.0$ cm, a breadth $b = 2.0$ cm, and with a thickness $w = 0.0254$ cm. System II channel dimensions are $V^0 = 4.25$ mL, $L = 90.0$ cm, $b = 1.9$ cm, and $w = 0.0254$ cm. The channel for system III was cut out of a 0.0127 cm thick Mylar spacer (thus $w = 0.0127$ cm) and has $V^0 = 1.19$ mL, $b = 1.0$ cm, and $L = 90.0$ cm. An Isochrom LC pump from Spectra-Physics (San Jose, CA) was used to pump the carrier liquid through the channel. The response from a UV detector (model 153 from Beckmann Instruments, Fullerton, CA) working at 254 nm was recorded and processed by FFFractionation software. The rotor radius representing the distance between the channel and axis of rotation in these systems is 15.1 cm.

SdFFF system IV, with the same basic features as the model S101 SdFFF units described above, is a research device used in the Field-Flow Fractionation Research Center laboratories. The channel dimensions of this system are $V^0 = 4.50$ mL, $L = 90.5$ cm, and $w = 0.0254$ cm. The channel has a single inlet but it has two outlets to allow stream splitting in order to enhance the detector signal (*34, 35*). The rotor radius for this SdFFF system is 15.3 cm. A Minipuls II peristaltic pump (Gilson, Madison, WI), a Beckmann model 153 UV detector working at 254 nm, and a strip chart recorder (Houston Instruments, Austin, TX) were used with system IV.

The carrier liquid was deionized and distilled water with 0.05% (*w/v*) sodium dodecyl sulfate and 0.01% (*w/v*) sodium azide both from Sigma (St. Louis, MO). The carrier liquid was purged with helium for degassing before pumping through the channel.

Scanning electron microscopy (SEM) was carried out with a Hitachi S-450 scanning electron microscope (Tokyo, Japan). For this, the sample was transferred onto a 0.1 μm pore size Nuclepore filter. The filter with the sample was then mounted on an SEM stub, gold coated with a Technics Hummer III sputter coater (Alexandria, VA), and subsequently examined under a 15 KV accelerating voltage.

The polystyrene (PS) latex standards and the polymethylmethacrylate (PMMA) latex beads used in this work were obtained from Seradyn Diagnostics (Indianapolis, IN). The other latex materials were obtained from various industrial sources.

Results and Discussion

Characterization of Narrow and Broad Latex Distributions. Figure 2 shows two fractograms of six narrowly distributed PS beads ranging from 0.2 to 0.9 μm diameter obtained by power programmed SdFFF under two sets of experimental conditions: (a) SdFFF system I, $G_0 = 675.9$ gravities (2000 rpm), $t_1 = 3$ min, $t_a = -24$ min, stop-flow time $t_{sf} = 6$ min, and flowrate $\dot{V} = 6.84$ mL/min; (b) SdFFF system II, $G_0 = 380.2$ gravities (1500 rpm), $t_1 = 13$ min, $t_a = -104$ min, $t_{sf} = 12$ min, and $\dot{V} = 6.37$ mL/min. These two fractograms demonstrate that experimental conditions can be varied flexibly to achieve particle separation at different resolution levels and in different run times. In fractogram *a*, although all PS beads are represented by peak maxima, none of the peaks is baseline resolved. On the other hand, the resolution level between adjacent peaks shown in fractogram *b* is more than adequate for baseline resolution. The difference in resolution levels between fractograms *a* and *b* in Figure 2 can be explained by equation 11. By substituting the above operating parameters into this equation, we find for fractogram *a* that $F_m = 0.62$ ($F_d = 1.86$) and for fractogram

b that $F_m = 3.9$ ($F_d = 11.8$). Thus the intrinsic resolving power (fractionating power) in fractogram b is 6.3 times higher than that in fractogram a. If we examine these calculations to see how the sixfold increase in fractionating power for fractogram b arises, we observe that the unequal values of G_0 and t^0 (inversely proportional to $\dot{V}$) contribute little to the difference in fractionating powers between fractograms a and b. The primary gain in F_m and F_d in fractogram b relative to fractogram a arises from the 4.3-fold increase in the time constant t_1 (and the corresponding increase in t_a) which, according to equation 11, gives a 7.1-fold gain in fractionating power. These calculations clearly rationalize the different resolution levels observed in the two fractograms in Figure 2 but, more importantly, they illustrate the important relationship between theory and experiment that can be utilized to choose experimental conditions such that they will yield the level of resolution desired by the operator. A complete description of experimental design is beyond the scope of this report.

In general, higher resolution is compromised by a shorter analysis time. The resolution level is enhanced (and the run time lengthened) by an increase in field strength at a constant carrier flowrate, a smaller flowrate at a constant field strength (consistent with equation 6), and, when field programming is used as in Figure 2, longer programs as dictated by larger time constants such as t_a and t_1.

The extraordinary resolving power of SdFFF was demonstrated recently by the separation of fused and normal doublets from one another and from the singlets in an aggregated 0.586 μm polymethylmethacrylate latex population (*19*). The mass of the fused doublets, as deduced form the SdFFF results, was about 10% less than that of the normal doublets (where the two spheres have minimal contact area). Separation of the two doublets and the measurement provided by SdFFF on their relative mass (e.g., through equation 4) not only highlighted their dual existence but served to delimit the mechanism by which they formed, providing best consistency with a two-stage growth mechanism (*19*).

High resolution can similarly be obtained with different sizes of PS latex beads as shown in Figure 3 where the separation of nominal 0.204 μm and 0.225 μm PS beads was achieved using SdFFF system I at 2000 rpm, a flowrate of 0.297 mL/min, and a stop-flow time of 10 min. These beads eluted as separate peaks with a resolution of 0.92. (The apparent fractionating powers are $F_m = 3.1$ and $F_d = 9.4$; the actual values are somewhat higher because the observed resolution is reduced by polydispersity.) The resolution level could be further enhanced if the separation was carried out with a slower flowrate using the same field strength, but the gain is limited by the finite polydispersity of the two PS latex samples.

Narrow PS samples do not represent real-world industrial latex samples which usually have broad particle size distributions. The ability of SdFFF to characterize broad samples is shown in Figure 4. The fractogram shown in Figure 4(a) is for an epoxy-acrylic dispersion and was obtained using SdFFF system I and power-pro-grammed field conditions with $G_0 = 169.0$ gravities (1000 rpm), $t_1 = 10$ min, $t_a = -80$ min, $t_{sf} = 10$ min, and $\dot{V} = 1.28$ mL/min. These conditions yield a fractionating power of $F_d = 2.0$ (from equations 11 and 12), a value adequate for this broad distribution. The expected sequential elution of small particles followed by larger particles (see equation 4) is confirmed by the inserted SEM micrographs of the original sample and of the three fractions collected at different elution times. The particle size distribution derived from the fractogram (without applying light scattering correction) is shown in Figure 4(b). For obtaining the PSD we used $\Delta\rho = 0.10$ g/mL. The mean particle diameter and the standard deviation were found to be 0.282 ± 0.107 μm. Our assumption for particle density yields diameters consistent (but not in exact agreement) with the sizes of particles shown in micrographs of cuts #1, #2 and #3. Specifically, the particle diameters of 0.187, 0.328, and 0.461 μm obtained by SdFFF (assuming $\Delta\rho = 0.10$ g/mL) correspond to diameters 0.184 ± 0.027, 0.299 ± 0.036, and 0.398 ± 0.030 μm obtained by SEM. (Since SdFFF provides slightly higher particle diameters

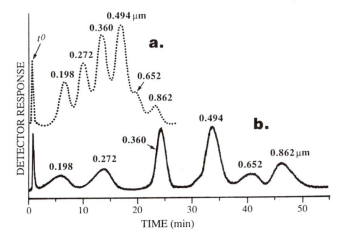

Figure 2. Contrasting separation of six narrow PS standards by power programmed SdFFF under two different sets of experimental conditions (see text).

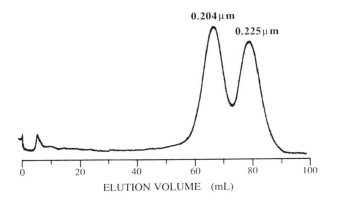

Figure 3. The extraordinary resolving power of SdFFF is demonstrated by the separation of nominal 0.204 and 0.225 μm PS beads, differing by only 9.8% in diameter.

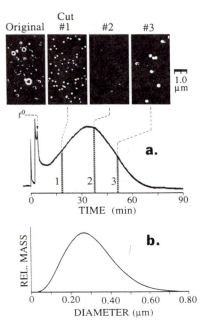

Figure 4. (a) Fractogram and (b) particle size distribution of a broad epoxy-acrylic latex sample. Micrographs of the original sample and three different fractions collected at different elution times are shown with the fractogram.

than SEM, the agreement between these results could be improved by assuming that the particle density difference $\Delta\rho$ is ~0.11 rather than 0.10 g/mL.) It is evident from the micrograph of the original sample and from the particle size distribution obtained by SdFFF that the sample population has a continuous but broad distribution.

Since both of the surfaces constituting the walls of the SdFFF channels used here are highly polished, either can serve as the accumulation wall depending on whether the sample particles float or sink through the carrier liquid. The polystyrene and epoxy-acrylic latex particles described above are both denser than water and therefore accumulate at the outer wall in dilute aqueous carriers (see Figure 1(a)). However polybutadiene particles, having a density of 0.92 g/mL, will accumulate at the inner wall as shown in Figure 1(b).

A fractogram of a broad polybutadiene sample obtained by SdFFF system III is shown in Figure 5(a). The experimental conditions were as follows: power-programmed field conditions with G_0 = 675.9 gravities (2000 rpm), t_1 = 10 min, t_a = -80 min, t_{sf} = 10 min, and $\dot{V}$ = 0.50 mL/min. The fractionating power is given by F_d = 6.1. The multimodal fractogram indicates that the particle distribution is rather complex in this sample. The particle size distribution curve derived from this fractogram and shown in Figure 5(b) is indeed quite broad with three distinct modes at 0.135, 0.272 and 0.457 µm. Note that the clear discernment between these modes by any technique requires good resolution, preferably F_d > 3. By using conditions that give F_d = 6.1, the three modes shown in Figure 5 are clearly distinguished and accurately characterized.

Characterization of Latex Aggregates. The high resolving power of SdFFF for separating latex aggregates composed of small numbers of nominal 0.230 µm PMMA beads is shown in Figure 6. SdFFF system IV was used to carry out the separation of these aggregates. Experimental conditions were as follows: constant field at 57.6 gravities (580 rpm), t_{sf} = 15 min, and $\dot{V}$ = 1.15 mL/min. The successive clusters, each having one more sphere than its predecessor, are found to elute as regularly spaced separate peaks as predicted by the center term of equation 4. SEM is used here as a complementary technique to verify the aggregation number (number of spheres) composing different aggregated clusters eluting in sequence.

Various examples of the aggregates composed of both PS and other PMMA latex beads was presented elsewhere (*19*). From trace to extensive levels of latex aggregation can be tracked by SdFFF (*18*). An approach for the determination of the polydispersity of latex aggregates from that of the elementary particles was outlined and experimentally verified (*17*).

Resolution as well as retention characteristics, which are influenced by the carrier flowrate, polydispersity of elementary particles, and the applied field strength, were examined in our earlier work (*16, 18*). Anomalies due to steric effects were shown to be significant for higher order latex aggregates as predicted by equation 13. Thus the spacing between successive peaks past the singlet is found to decrease continuously as the aggregation number increases.

The change in elution profiles arising from sample aging and attendant aggregation (manifest by a decrease in the singlet population and an increase in the amounts of higher order clusters with time) was observed for a nominal 0.325 µm PMMA latex. The evolution of cluster populations due to sample aging and aggregation can be followed in detail by the comparison of successive particle size distributions (*18*). This approach can be useful for monitoring aggregation in both narrow and broad latex dispersions. Aggregate distintegration as well as formation can be surveyed. For example, the breakup of 0.299 µm PMMA latex aggregates by sonication was monitored by SdFFF (*19*). Studies on cluster formation due to the addition of selective surfactants (*19, 36*), proteins (*37*), and antigens (unpublished results) to a population of uniform latex beads were also carried out.

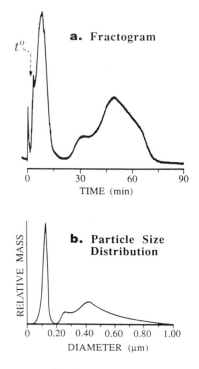

Figure 5.   (a) Fractogram of a broad polybutadiene latex sample and (b) multimodal particle size distribution derived from the fractogram shown in (a).

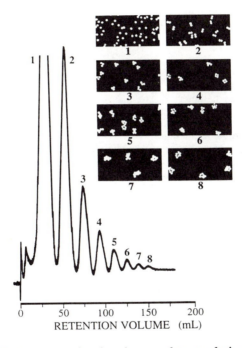

RETENTION VOLUME (mL)

Figure 6. SdFFF fractogram showing the complete resolution of eight sizes of PMMA aggregates. Electron micrograph of latex beads collected from each peak show that pure populations of different sized clusters are isolated by SdFFF. (Results from reference 2.)

Conclusions

The capability of SdFFF to achieve the separation and characterization both of narrow and broad distributions of polymeric latex beads, along with aggregated latex populations, and to fractionate particles either denser or less dense than the carrier liquid, is demonstrated here. Examples are provided using both latex standards and industrial latexes. Equations are given to help select experimental conditions needed to achieve desired resolution levels. It is also shown that particle size distributions, even for complex multimodal particle distributions, can be derived from the SdFFF elution profiles. SdFFF and related FFF techniques are unique in that the validity of results obtained by these methods can be verified, and further characterization realized, by the collection of particles at specified elution times and their examination by complementary techniques such as SEM.

Legend of Symbols

b	channel breadth
d	particle diameter
d_1	singlet particle diameter
d_n	effective spherical diameter of n aggregated particles
G	centrifugal acceleration
G_0	initial centrifugal acceleration
H_p	polydispersity contribution to plate height
k	Boltzmann constant
ℓ	mean particle elevation (layer thickness) above accumulation wall
L	channel length
m	particle mass
n	aggregation number
R	retention ratio
R_s	resolution
t	elapsed time
t^0	void time
t_1	time lag before field decrease
t_a	time constant for power-programming
t_{sf}	stop-flow time
T	absolute temperature
$<v>$	mean linear flow velocity
$\dot{V}$	channel flowrate
V^0	void volume
V_r	retention volume
w	channel thickness

Greek symbols

γ	steric correction factor
Δd	size difference of two particles
Δm	mass difference of two particles
$\Delta \rho$	difference in density between particle and carrier liquid
λ	dimensionless retention parameter
ρ_p	particle diameter
σ_d	standard deviation in particle diameter

Acknowledgments

This work was supported by Grant CHE-9102321 from the National Science Foundation.

Literature Cited

1. Giddings, J. C.; Yang, F. J. F.; Myers, M. N. *Anal. Chem.*, **1974**, *46*, 1917-1924.
2. Giddings, J. C. *Chem. Eng. News* **1988**, *66* (October 10), 34-45.
3. Caldwell, K. D. *Anal. Chem.* **1988**, *60*, 959A-971A.
4. Kirkland, J. J.; Yau, W. W.; Doerner, W. A. *Anal. Chem.* **1980**, *52*, 1944-1954.
5. Giddings, J. C.; Myers, M. N.; Moon, M. H.; Barman, B. N. In *Particle Size Distribution II: Assessment and Characterization*; Provder, T., Ed.; ACS Symp. Series No. 472; American Chemical Society: Washington, DC, 1991; pp 198-216.
6. Giddings, J. C.; Myers, M. N. *Sep. Sci. Technol.* **1978**, *13*, 637-645.
7. Koch, T.; Giddings, J. C. *Anal. Chem.* **1986**, *58*, 994-997.
8. Bangs, L. B. *Uniform Latex Particles*; Seradyn Inc.: Indianapolis, IN, 1984.
9. McDonald, P. J. *J. Water Borne Coat.* **1982**, *5*, 11-23.
10. Del Gatto, J. V., Ed. *Materials and Compounding Ingredients for Rubber*; Rubber/Automotive Publications: New York, NY, 1970.
11. Echte, A. In *Rubber-Toughened Plastics*; Riew, C. K., Ed.; Adv. Chem. Series No. 222; American Chemical Society: Washington, DC, 1989; pp. 15-64.
12. Provder, T., Ed.; *Particle Size Distribution II: Assessment and Characterization*, ACS Symp. Series No. 472; American Chemical Society: Washington, DC, 1991.
13. Family, F.; Landau, D. P., Eds.; *Kinetics of Aggregation and Gelation*, Elsevier: New York, 1984.
14. Allen, T. *Particle Size Measurement*; 3rd ed.; Chapman and Hall: London, 1981.
15. Caldwell, K. D.; Nguyen, T. T.; Giddings, J. C.; Mazzone, H. M. *J. Virol. Methods* **1980**, *1*, 241-256.
16. Jones, H. K.; Barman, B. N.; Giddings, J. C. *J. Chromatogr.* **1989**, *455*, 1-15.
17. Giddings, J. C.; Barman, B. N.; Li, H. *J. Colloid Interface Sci.* **1989**, *132*, 554-565.
18. Barman, B. N.; Giddings, J. C. In *Particle Size Distribution II: Assessment and Characterization*; Provder, T., Ed.; ACS Symp. Series No. 472; American Chemical Society: Washington, DC, 1991; pp 217-228.
19. Barman, B. N.; Giddings, J. C. *Langmuir* **1992**, *8*, 51-58.
20. Barman, B. N.; Giddings, J. C. *Polym. Mater. Sci. Eng.*, **65**, 61-62 (1991).
21. Giddings, J. C.; Karaiskakis, G.; Caldwell, K. D. *Sep. Sci. Technol.* **1981**, *16*. 607-618.
22. Giddings, J. C.; Caldwell, K. D.; Jones, H. K. In *Particle Size Distribution: Assessment and Characterization*; Provder, T., Ed.; ACS Symp. Series No. 332; American Chemical Society: Washington, DC, 1987; pp 215-230.
23. Nagy, D. J. *Anal. Chem.*, *61*, 1934-1937 (1989).
24. Kirkland, J. J.; Yau, W. W. *Anal. Chem.* **1983**, *55*, 2165-2170.
25. Giddings, J. C.; Yoon, Y. H.; Caldwell, K. D.; Myers, M. N.; Hovingh, M. E. *Sep. Sci.* **1975**, *10*, 447-460.
26. Giddings, J. C.; Williams, P. S.; Beckett, R. *Anal. Chem.* **1987**, *59*, 28-37.
27. Martin, M.; Giddings, J. C. *J. Phys. Chem.* **1981**, *85*, 727-733.
28. Williams, P. S.; Giddings, J. C. *Anal. Chem.* **1987**, *59*, 2038-2044.

29. Yang, F. J. F.; Myers, M. N.; Giddings, J. C. *Anal. Chem.* **1974**, *46*, 1924-1930.
30. Williams, P. S.; Giddings, J. C.; Beckett, R. *J. Liq. Chromatogr.* **1987**, *10*, 1961-1998.
31. Yau, W. W.; Kirkland, J. J. *Sep. Sci. Technol.* **1981**, *16*, 577-605.
32. Williams, P. S.; Giddings, J. C. *J. Chromatogr.* **1991**, *550*, 787-797.
33. Lee, S.; Giddings, J. C. *Anal. Chem.* **1988**, *60*, 2328-2333.
34. Giddings, J. C.; Lin, H. C.; Caldwell, K. D.; Myers, M. N. *Sep. Sci. Technol.* **1983**, *18*, 293-306.
35. Jones, H. K.; Phelan, K.; Myers, M. N.; Giddings, J. C. *J. Colloid Interface Sci.* **1987**, *120*, 140-152.
36. Li, J.; Caldwell, K. D.; Tan, J. S. In *Particle Size Distribution II: Assessment and Characterization*; Provder, T., Ed.; ACS Symp. Series No. 472; American Chemical Society: Washington, DC, 1991; pp 247-262.
37. Beckett, R.; Ho, J.; Jiang, Y.; Giddings, J. C. *Langmuir* **1991**, *7*, 2040-2047.

RECEIVED July 6, 1992

Chapter 4

Polymer Separation and Molecular-Weight Distribution by Thermal Field-Flow Fractionation

Marcus N. Myers, Peter Chen, and J. Calvin Giddings

Field-Flow Fractionation Research Center, Department of Chemistry, University of Utah, Salt Lake City, UT 84112

Thermal field-flow fractionation, an FFF method in which the driving force is generated by a temperature gradient, has proven highly effective in the separation and analysis of lipophilic synthetic polymers. Thermal FFF is more flexible than SEC (GPC) and, because the FFF channel has no packing, there is less risk of shear degradation, clogging, and system degradation.

    Here theoretical guidelines and examples are given to show how run time and fractionating power can be flexibly manipulated toward desired goals through changes in flowrate and temperature drop. The calibration procedure for polymer molecular weight analysis is described. Molecular weight distributions are obtained both for polystyrene and polyethylene samples. The polyethylene analysis requires modified equipment capable of operating at high cold wall temperatures.

Field-flow fractionation (FFF), in its various forms, is applicable to the separation and characterization of virtually all categories of macromolecular, colloidal, and particulate matter (*1-4*). Different subtechniques of FFF have been found preferable for different applications. For particle size analysis, sedimentation FFF and flow FFF are most effective (see chapters 2 and 3). For the analysis of lipophilic polymers, thermal FFF has emerged as the FFF subtechnique of choice (*5-10*). (Another promising but less developed candidate for polymer analysis is flow FFF which has been applied both to polymers soluble in water (*11, 12*) and to those soluble in organic solvents (*13*).) Thermal FFF is the FFF subtechnique in which the driving force, directed perpendicular to the flow/separation axis, is generated by a temperature gradient and is based on the phenomenon of thermal diffusion (*5-8*). The temperature gradient is generated by sandwiching a thin (~0.01 cm) elongated rectangular channel between hot and cold metal bars (see Figure 1). Temperature gradients up to 10,000°C/cm are produced. These gradients are sufficient, despite the inherent weakness of thermal diffusion as a driving force, to drive the transport underlying polymer separation in thermal FFF. The thermal FFF method has been found applicable to virtually all polymers examined including polystyrene (PS), polymethylmethacrylate (PMMA), polyisoprene, polysulfone, polycarbonate, nitrocellulose, and polybutadiene. A high temperature thermal FFF system (*14*) has been applied to polyolefins including polyethylene (see later). Various copolymers have been subjected to thermal FFF analysis as well (15). (See chapter 5 in this volume by Schimpf, Wheeler, and Romeo for further information on copolymer studies.)

Thermal FFF and size exclusion chromatography (SEC) are both flow/elution techniques that conveniently separate polymers according to differences in chain length or molecular weight. The single most striking feature of thermal FFF compared to SEC is that the fundamental process underlying separation is driven by a temperature gradient rather than by partitioning into a porous support (5-7). This leads to several differences in operation and performance. Since the temperature gradient can be immediately and precisely controlled but the support pore size in an SEC column cannot, thermal FFF has much greater flexibility. The temperature drop ΔT in a thermal FFF channel can be adjusted to a high value (e.g., 100°C) for low molecular weight (M) polymers or a low value (e.g., 10°C) for high M polymers. If the M range is large, the thermal FFF system can be programmed starting with a high ΔT for the low M components that is then programmed downward to some suitable lower ΔT to complete the separation of high M components (5, 16). The range in ΔTs is adjusted to accommodate the M range of the polymer sample. An example is shown in Figure 2 using PS standards. For the 100-fold range in polymer M (from 35,000 to 3,800,000 Daltons), separation can be achieved with an initial ΔT of 80°C, which is subsequently decayed according to the mathematical form of power programming (5, 17), leading to the rapid elution of all six components. The separation is complete in 20 minutes.

The most unusual feature of Figure 2 from the viewpoint of SEC practitioners is that the elution sequence proceeds from low to high M. This feature by itself has very little bearing on molecular weight analysis since calibration can be established and molecular weight distribution curves obtained using either sequence and thus either method.

Other differences relative to SEC arise because of the different thermal FFF retention mechanism and channel structure. Since differential retention is induced by the temperature gradient between flat channel walls and not by the pores of the granular packing, the FFF channel is an open structure with no fixed particles obstructing the flow. Thus the flow pattern in thermal FFF lacks the tortuosity characteristic of SEC flow. Consequently shear-sensitive molecules are subjected to less disruptive conditions in the thermal FFF channel.

Shear sensitivity is an important consideration for many high molecular weight polymers (18, 19). Thermal FFF has been used successfully on many occasions (see Figure 2) to fractionate polymers with $M > 10^6$ Daltons and in one case for $M > 10^7$ Daltons (20). On the other hand, thermal FFF has difficulty in resolving polymers below $M = 5000$-$10,000$ without strong measures to increase ΔT. This is due to the weakness of thermal diffusion effects for low MW polymers. However, with ΔT increased to 150°C, polymers down to 1000 MW can be fractionated (21).

One of the unique characteristics of thermal FFF is that retention depends not only on MW but also on the chemical composition of the polymer. This "chemical" differentiation is due to the dependence of the underlying thermal diffusion process on polymer (and solvent) composition (22). This effect can likely be used to determine compositional distributions in copolymers and blends (15). More detail is provided in chapter 5 by Schimpf et al.

The thermal FFF channel is more robust than an SEC column on several counts. First, since the FFF channel is open, it has little tendency to clog, even in the presence of particles (including gels) up to 75 or 100 μm diameter, which is the typical range for channel thickness. Second, there is no packing in the channel, whose degradation, particularly at high temperatures, can lead to shifts in retention and thus in apparent molecular weight. In this respect thermal FFF is particularly promising for high temperature polyolefin analysis (see below).

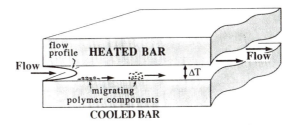

Figure 1. Short section of thermal FFF channel system showing the differential migration of two polymer components. The degree to which the polymer molecules are driven toward the wall and thus retained depends upon the temperature drop ΔT, which is a controllable parameter.

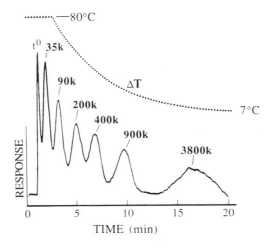

Figure 2. Fractogram showing the separation of polystyrene (PS) standards of the indicated molecular weights in THF using power programmed thermal FFF with a flowrate of 0.5 mL/min and a t_1/t^0 ratio of 2.0. The programmed temperature drop (ΔT) profile is shown.

Theory

The theory of FFF generally (2, 23), and the theory specific to thermal FFF (24, 25), have been reported elsewhere. Here we summarize only a few essential elements of theory to aid understanding. Some emphasis is placed, however, on the theory showing how run time and fractionating power can be flexibly controlled through changes in flowrate, temperature drop, and programming parameters.

The retention of polymers in thermal FFF is induced by the temperature gradient, which drives polymer molecules toward the cold wall of the channel. As the polymer components approach the wall, their velocity of displacement by the solvent stream is reduced by frictional drag at the wall. This frictional drag is responsible for a flow profile that is approximately parabolic in shape as shown in Figure 1. The closer any given polymer molecule approaches the wall, the slower its displacement velocity due to its entrainment in increasingly slow streamlines in the parabolic flow profile. The effectiveness of thermal FFF is based on the fact that high molecular weight polymers are driven closer to the wall and thus retained more than low molecular weight polymers. This differential retention and the associated differences in migration velocities lead to separation as illustrated in Figure 1.

Retention in FFF is specified by the *retention ratio R*, which is the ratio of the migration velocity of a specified polymer fraction to the mean velocity <v> of the carrier solvent through the channel. For thermal FFF, R is given approximately by (26)

$$R \cong 6\lambda \cong \frac{6D}{D_T \Delta T} \tag{1}$$

where λ is the retention parameter, D is the ordinary diffusion coefficient of the polymer, D_T is the thermal diffusion coefficient, and ΔT is the difference in temperature between hot and cold walls.

It has been found that D_T is independent of chain length and branching (22, 27). This is in contrast to D, which is known to have a molecular weight dependence of the following form for random coil polymers (28, 29)

$$D = AM^{-b} \tag{2}$$

where A is a constant dependent on the polymer-solvent system and temperature and b is a more universal constant of magnitude $\cong 0.6$. In view of this relationship and the lack of (or very weak) dependence of D_T on M, we can write

$$\log \frac{D}{D_T} = \log \lambda \Delta T = \log \phi - n \log M \tag{3}$$

where ϕ and $n \cong b$ can be considered as empirical constants. This equation serves as the basis for our calibration.

For constant field runs (carried out at constant ΔT), the retention time t_r is related to the void time t^0 (the passage time of solvent carrier or of a nonretained peak) by

$$t_r = t^0/R \tag{4}$$

Time t^0 is related to the void volume V^0 of the channel and the volumetric flowrate $\dot{V}$ through the expression

$$t^0 = V^0/\dot{V} \tag{5}$$

When equations 1 and 5 are used in conjunction with equation 4, we get (5)

$$t_r \cong \frac{D_T \, \Delta T \, t^0}{6D} = \frac{D_T \, \Delta T \, V^0}{6D\dot{V}} \tag{6}$$

which shows that the retention time is approximately proportional to the field strength ΔT and inversely proportional to the flowrate $\dot{V}$.

For field programming, the retention ratio R varies continuously as ΔT is programmed. Because R varies, equation 4 cannot be used; it is replaced by the integral form (16)

$$t^0 = \int_0^{t_r} R \, dt \tag{7}$$

In order to integrate $R \, dt$, the time dependence of R must be specified. However, R simply follows the programmed changes in ΔT as specified by equation 1. A ΔT program designed to provide a constant fractionating power (see below) across the entire molecular weight range of the polymer components is given by the power programming function (5)

$$\Delta T = \Delta T_0 \left(\frac{3t_1}{t + 2t_1} \right)^2 \tag{8}$$

which applies when time t exceeds the lag time t_1, the period in which ΔT is held constant at its initial value ΔT_0. With the program given by equation 8, the retention time for well retained components becomes (5)

$$t_r = t^0 \left[\frac{D_T \, \Delta T_0}{2D} \left(\frac{3t_1}{t^0} \right)^2 \right]^{1/3} - 2t_1 \tag{9}$$

This expression replaces equation 6, which is only applicable under constant field conditions.

Resolution in thermal FFF is measured by the mass-based fractionating power defined by (30, 31)

$$F_M = \frac{R_s}{\delta M/M} \tag{10}$$

where R_s is the resolution of two polymer components whose average molecular weight is M and difference in molecular weight is δM. (Here $\delta M/M$ is assumed to be small, which means that the two components are close lying.) For well-retained components ($R \ll 1$), we can use the theory of nonequilibrium plate height and resolution in FFF (24) to get the following fractionating power for a constant ΔT run

$$
F_M = \frac{n}{8Dw} \left[\frac{(D_T \Delta T)^3 V^0}{6V} \right]^{1/2} = \frac{nD_T \Delta T}{8Dw} \left(\frac{D_T \Delta T \, t^0}{6} \right)^{1/2}
\tag{11}
$$

By contrast, the limiting fractionating power applicable to the power program defined by equation 8 is given by (5)

$$
F_M = \frac{3nt_1}{8wt^0} (2D_T \Delta T_0 \, t^0)^{1/2}
\tag{12}
$$

We note that for constant field operation F_M is inversely proportional to D and is thus proportional to M^b as shown by equation 2. However, the fractionating power for power programming, as expressed by equation 12, is independent of molecular weight, a constancy that distinguishes this form of programming from others (32).

Experimental Section

Conventional Thermal FFF. The channel system employed in these studies has the same basic structure as the Model T100 thermal FFF system from FFFractionation, Inc. (Salt Lake City, UT). Two channels, differing in thickness, were assembled using this system. The channel volume in each case is cut from a thin (76 or 114 μm) polyester (Mylar) spacer sandwiched between two highly polished chrome-plated bars of electrolytic grade copper. The resulting channel dimensions (tip-to-tip length × breadth × thickness) are 46.3 cm × 2.0 cm × 0.0076 cm for the thinner channel and 46.3 cm × 2.0 cm × 0.0114 cm for the thicker channel. The void volumes based on geometry and on void peak measurements were 0.685 mL and 0.740 mL, respectively, for the thin channel and 1.045 mL and 1.100 mL for the thicker channel. For the void peak measurement, the dead volume of the tubing in the detector was subtracted from the measured volume of the void peak; these volumes were 8% and 5% of the channel volume for the thin and thick channels, respectively.

The upper bar was heated by four cartridge heaters of 1500 watts each. The temperature of the hot wall was controlled by computer-activated solid-state relays. Two thermistor probes from Thermometrix (Edison, NJ) were placed in two small holes drilled into the copper bars and approaching the hot and cold surfaces, respectively. In this way the hot wall temperature, the cold wall temperature, and the temperature difference ΔT could be monitored continuously during the run.

The carrier solvent (THF) was delivered to the channel by an Isochrom LC Model pump from SpectraPhysics (San Jose, CA). The flowrates were checked using a stopwatch and a buret. Samples consisted of 10 μL volumes of 0.1% solutions of polystyrene polymers (described in Table I) introduced by microsyringe into a Rhodyne Model 7125 injector (Cotati, CA). Following injection, a 30 second stopflow period was used for sample relaxation. The eluting polymer fractions were detected by using an Altex Model 153 UV detector (Beckmann Instruments, San Ramon, CA) operating at 254 nm wavelength. The data were recorded both on a Servogor Model 120 chart recorder (BBC Goerz Metrawatt, Vienna) and an 8088XT computer.

Table I. Linear Polystyrene Standards Used in This Study

Molecular Weight and Lot No.	Supplier	Polydispersity
35,000/LA16965	Supelco	≤1.06
47,500/LA20775	Supelco	≤1.06
90,000/PS50522	Pressure Chemical Co.	≤1.06
200,000/PS50912	Pressure Chemical Co.	≤1.06
200,700/--	Supelco	≤1.05
400,000/PS00507	Supelco	≤1.06
575,000/PS30121	Pressure Chemical Co.	≤1.06
900,000/PS80323	Pressure Chemical Co.	≤1.06
1,050,000/---	NIST	

Operation was carried out both at constant ΔT and with power programmed ΔT. (Power programming is a unique form of programming providing a constant fractionating power as described in the Theoretical Section.) For power programmed operation, the initial ΔT was usually set at 100°C. Other parameters are reported with the experimental data.

High Temperature (Polyethylene) Thermal FFF. In order to fractionate polyethylene and other polymers whose temperatures must be kept well above ambient, a modified apparatus is necessary. To this end, a slightly modified channel system (channel dimensions 46.3 cm × 2.0 cm × 0.0127 cm) with a polyimide spacer was placed in the column compartment of a Model 150-CV Gel Permeation Chromatograph (Waters Division of Millipore, Milford, MA). The modified channel system was thoroughly insulated from the remaining apparatus by wrapping in fiberglass. Heat was removed from the "cold" wall (held at 135 ± 1°C) by pumping deionized water at ~150 mL/min through the cooling conduits in the channel using an FMI QD1 pump (Fluid Metering, Inc., Oyster Bay, NY). The water was preheated to ~120°C by passing through copper tubing wrapped around a 500 watt cartridge heater (Watlow, St. Louis, MO). Heat was removed by vaporization of the cooling water. The boiling point was controlled by adjusting the coolant pressure, achieved by using a needle valve where the coolant exits the channel system. As has been shown in a previous study, removal of heat by coolant vaporization provides a constant cold wall temperature and a simple mechanism (pressure adjustment) for altering the cold wall temperature (*14*).

The hot wall of the thermal FFF channel was controlled by a dedicated computer built in house. The temperature of the hot and cold walls was monitored using J-type thermocouples and Omega Model 650 thermometers (Omega Engineering, Stamford, CT).

The polyolefin samples were dissolved at a concentration of 0.05% in Omnisolve 1,2,4-trichlorobenzene (EM Science, Biggtown, NJ) containing 0.1% Irganox 1010 (Ciba-Geigy, Hawthorne, NY) as an antioxidant. The samples were heated at 170°C in an oven for several hours until completely dissolved. Heating time depended upon the molecular weight of the sample. The sample was then transfered to the heated carousel of the Model 150 CV. The autosampler chamber of the Model 150 CV was kept at 145°C to maintain sample solubility during the analysis.

The polyethylene samples were obtained from NIST: The $\overline{M}_w$ = 119.6k sample is NIST standard 1484 while the other two ($\overline{M}_w$ = 158k and 200k), given to us several years ago, are not regularly offered by NIST.

Flexibility of Thermal FFF

We have already noted that thermal FFF has great flexibility in meeting experimental demands on resolution, analysis speed, and molecular weight range. Much of this flexibility arises because of the ready controllability of two parameters that strongly influence retention time and fractionating power (or resolution): temperature drop ΔT and channel flowrate $\dot{V}$. Their importance for constant ΔT runs is shown by equations 6 and 11. Both parameters can be varied from run to run to adapt to special needs. In addition, both parameters can be programmed to accommodate broad MW samples. However, only ΔT is commonly programmed at the present time. Equations 9 and 11 demonstrate the importance of $\dot{V}$ (expressed in terms of t^0 in equation 9, the connection with $\dot{V}$ being made through equation 5) and the initial ΔT (or ΔT_0) for programmed operation.

It is instructive to illustrate the influence of ΔT and $\dot{V}$ on experimental polymer separation and to show how these parameters can be manipulated to achieve different objectives. In Figure 3 we show the effects of changing ΔT (which in this case is not programmed) on the separation of three polystyrene standards in the 76 μm thick channel. The flowrate $\dot{V}$ is held constant at 0.2 mL/min. We observe that as ΔT is increased from 50°C to 70°C, the retention time of all polymer components increases. This increase is predicted by equation 6. However, concomitant with the increase in retention time is a clear gain in polymer resolution, a gain predicted explicitly by equation 11. This gain becomes particularly valuable for some lower MW components that elute closer to the void (nonretained) peak.

With regard to adjustments in flowrate, just like those in ΔT described above, FFF techniques exhibit a tradeoff between speed and resolution. Greater speed can be gained by increasing $\dot{V}$ (equation 6) with some sacrifice in resolution (equation 11) or, conversely, resolution can be gained by reducing $\dot{V}$ for a longer run.

If analysis time, for example, is found to be excessive but there is resolution to spare, $\dot{V}$ can be increased. Figure 4 shows the fractogram obtained with a threefold gain in $\dot{V}$ (to 0.6 mL/min) relative to the fractogram produced (and first displayed in Figure 3) at 0.2 mL/min, both obtained at ΔT = 70°C in the 76 μm channel. The analysis time is inversely proportional to flowrate (see equation 6) so the threefold multiplication of $\dot{V}$ reduces the run time from 27 min to <9 min. Resolution is seen to be lost at the higher $\dot{V}$ (by a factor of $1/\sqrt{3}$ in accord with equation 11) but it is still quite satisfactory.

By and large, optimal conditions are those in which both ΔT and $\dot{V}$ are relatively high. Gains in fractionating power are more readily achieved by increasing ΔT than by decreasing $\dot{V}$ as shown clearly by equation 11. The increased retention times associated with a larger ΔT can be offset by a gain in $\dot{V}$ that is directly proportional to the gain in ΔT as shown by equation 6. This matter can be further clarified by solving for $\dot{V}$ from equation 6 and substituting this expression into equation 11. This combination produces

$$F_M = \frac{nD_T\,\Delta T}{8w}\left(\frac{t_r}{D}\right)^{1/2} = \frac{n\,\Delta T}{8w}\left(\frac{D_T\,t_r}{\lambda\,\Delta T}\right)^{1/2} \tag{13}$$

which shows that for a fixed retention time (obtained by adjusting $\dot{V}$) the fractionating power always increases with ΔT. However, fractionating power is lost gradually as one increases the flowrate and thus the run speed at constant ΔT. While equation 13

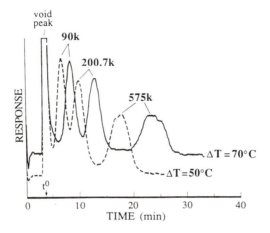

Figure 3.   Comparison of thermal FFF fractograms (76 μm channel) of polystyrene standards in THF at ΔTs of 50°C and 70°C and at a flowrate (V) of 0.2 mL/min.

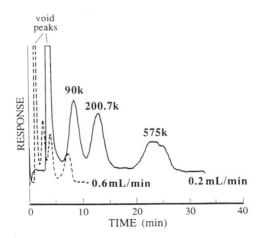

Figure 4.   Fractograms of three polystyrene standards obtained from the 76 μm thick channel with ΔT = 70°C at two different flowrates: 0.2 and 0.6 mL/min.

shows the benefit of maximizing ΔT, care must be taken because a high ΔT separation is more subject to disturbances by sample overloading than one at lower ΔT.

With regard to specific values of the fractionating power, the use of equation 11 (or equation 13) shows that F_M varies from 1.2 to 3.4 as one proceeds from $M = 90,000$ to 575,000 in the $\Delta T = 50°C$ fractogram of Figure 3. These values increase to 1.9 and 5.6, respectively, for the 70°C fractogram. When the flowrate is increased to 0.6 mL/min, as shown in Figure 4, the F_M range falls back to 1.1 and 3.2, respectively. The diffusion coefficients necessary for the above calculations were obtained from previous studies (22). We have not attempted to make a quantitative comparison between these values and the experimental results because sample polydispersity causes additional band broadening and thus causes an apparent (but not actual) loss of resolving power (33).

We have noted already that ΔT can be programmed to expand the molecular weight range covered. This was illustrated in Figure 2. However, changes in ΔT and $\dot{V}$ have the same underlying effects in programmed runs as for nonprogrammed operation. Altogether, programmed runs can be optimized both by adjustments in $\dot{V}$ and ΔT_0 and by changes in the time scale of the program as expressed by the ratio t_1/t^0. These parameters have been chosen to provide convenient speed and yet adequate resolution for the polymer standards separated in Figure 2. The fractionating power calculated from equation 12 for this run is 2.3.

Molecular Weight Distribution of Polystyrene

Software and an associated calibration procedure for molecular weight distribution analysis have been developed and will be reported in detail elsewhere. For polydisperse polystyrene samples, calibration is achieved using the observed retention times of a set of polystyrene standards. We first obtain D/D_T values from the observed retention times or ratios using a modified form of equation 1 that provides increased accuracy. We then obtain the desired calibration by plotting the logarithm of D/D_T against the logarithm of molecular weight as suggested by equation 3. A least squares straight-line fit to these data is then used for calibration purposes. Figure 5 shows an example of a calibrating fractogram obtained using three PS standards ($M = 47.5k, 200k, 900k$) and the corresponding calibration plot generated using the 76 μm thick channel with power programming and THF carrier. The conditions were $\Delta T_0 = 85°C$, $\dot{V} = 0.2$ mL/min, and $t_1 = 5$ min. The parameters determined from the calibration plot (see equation 3) are $\phi = 10640$ and $n = 0.6433$.

Figure 6 demonstrates the application of this calibration procedure to the high molecular weight NIST reference polystyrene 1479. For improved accuracy, calibration is generally carried out using the same experimental conditions (including flowrate, temperature drop, and programming conditions, if any) as applied to the sample itself. Thus the fractogram shown in Figure 6a was generated using the 76 μm thick channel operated under the same programmed conditions (which yield $F_M = 2.5$) as utilized for the calibration run of Figure 5. The molecular weight distribution of NIST 1479 obtained from the fractogram of Figure 6a using this calibration procedure and data is shown in Figure 6b. This distribution has weight and number average molecular weights of 1.10 and 1.02 million Daltons, respectively. The NIST value for $\overline{M}_w$ is 1.05 million, which agrees reasonably well with the thermal FFF value.

The above procedure can be readily applied to industrial polystyrenes. Figure 7a shows the fractograms ($F_M = 2.3$) generated by the 112 μm thick channel with $\dot{V} = 0.5$ mL/min, $\Delta T_0 = 100°C$, and $t_1 = 5$ min for three different polystyrene samples acquired from miscellaneous industrial foam products including two different foam shipping boxes (fractograms A and B) and a polystyrene beverage cup (C). (The void peak has been subtracted from these fractograms, as well as from those in Figures 5

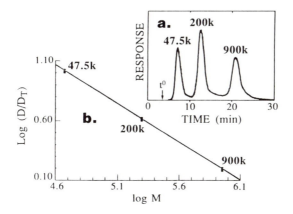

Figure 5. Fractogram (a) of mixture of three polystyrene standards and straight line calibration plot (b) of log D/D_T versus log molecular weight obtained from fractogram retention data.

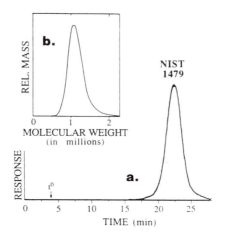

Figure 6. Thermal FFF fractogram (a) (same conditions as Figure 5) and molecular weight distribution (b) of NIST 1479 polystyrene using calibration plot of Figure 5.

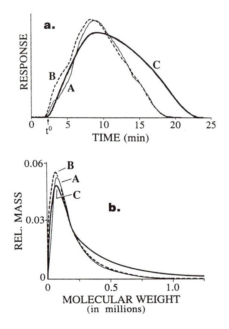

Figure 7. Fractograms (a) and molecular weight distribution curves (b) for three industrial polystyrene samples obtained from miscellaneous foam products.

and 6, in preparation for data processing.) The leading (low molecular weight) edge of these fractograms are all similar but sample 3 displays an elevated trailing edge corresponding to a relatively larger content of high molecular weight polymer. This observation is confirmed by the molecular weight distribution curves obtained (using in this case ϕ = 10,700 and n = 0.6500) from the three polystyrene fractograms and shown in Figure 7b. The three polystyrenes are found to have $\overline{M}_w$ values of 208k, 206k, and 332k, respectively. The respective number averages are 98k, 81k, and 125k.

High Temperature Thermal FFF of Polyethylene

Many polymers, including but not limited to the polyolefins, must be analyzed at high temperature. Polyethylene and polypropylene, for example, require a minimum temperature of around 130°C to ensure polymer solubility. The cold wall temperature in conventional thermal FFF, ~20°C, is clearly far too low for the analysis of these polymers. However, by altering the cooling system such that heat is removed from the cold wall by water (or other coolant) vaporization rather than by cold water flow, the cold wall temperature can be elevated to the desired level. This principle was demonstrated in earlier work done in our laboratory (*14*). More recently, we have inserted a thermal FFF channel into a Waters 150-CV instrument, replacing the normal SEC (GPC) column. The Waters instrument provides a capability for injection and detection at high temperatures. However, the guest channel requires independent temperature control, heat input, and heat removal components to function. These were described in the experimental section.

Preliminary results are promising. In Figure 8 we show the elution profiles (fractograms) of three polyethylene samples from NIST, each with a different reported $\overline{M}_w$ as shown. We observe that with increasing $\overline{M}_w$, the polymer profile elutes later, as expected.

In order to obtain quantitative molecular weight distributions for polyethylene, we must determine the calibration constants ϕ and n, but there are no narrow standards to be used for this purpose. Therefore we started by using calibration constants typical for polystyrene, obtained a molecular weight distribution for polyethylene based on these trial constants, calculated $\overline{M}_w$ values from the polyethylene distributions, adjusted the constants to produce $\overline{M}_w$ values in better agreement with reported values, and by a few iterations of this procedure arrived at the values ϕ = 1.25 × 10^6 and n = 1.1 for polytheylene. With these constants and the calibration procedure discussed above, the molecular weight distributions for polyethylene samples can be obtained. The distributions for the three NIST samples fractionated in Figure 8a are shown in Figure 8b. From these distributions we calculate weight averages of 119.6k, 154.5k, and 208.6k, number averages of 112.3k, 147.5k, and 196.6k, respectively, for the three polyethylenes of increasing molecular weight. The polydispersities (weight/number average) are 1.07, 1.05, and 1.06, respectively. For the lower molecular weight component, NIST 1484, the polydispersity is reported as ~1.19, a value considerably larger than the 1.07 found here. Based on previous studies (*34*), thermal FFF measurements tend to overestimate rather than underestimate polydispersity and we therefore think it likely that the true polydispersity is ≤1.07. (The one factor that would alter this conclusion would be the presence of a low molecular weight population not retained by thermal FFF and therefore not accounted for in the molecular weight distribution curves of Figure 8b.)

These preliminary results on the molecular weight distribution of polyethylene need considerable refinement and additional study to establish calibration constants that can be applied with confidence. Thus in Figure 8a the 200k polyethylene is retained longer than NIST 1484 (119.6k) in accordance with expectation, but the additional retention is more than expected based on other polymer studies. This explains why the

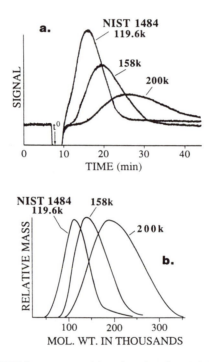

Figure 8. Thermal FFF fractograms (a) and molecular weight distributions (b) of three NIST polyethylene samples of the indicated $\overline{M}_w$ values run in trichlorobenzene at $\dot{V} = 0.1$ mL/min, $\Delta T = 70°C$, and a cold wall temperature of 135°C.

calibration constant $n = 1.1$ was needed to fit the data, whereas typically $n = 0.55$ to 0.75. This difference may reflect some fundamental property of polyethylene behavior (e.g., a D_T value dependent on M), an error in reported M values, or the possibility of sample overloading in our FFF runs. Once this anomaly is resolved, thermal FFF is likely to become a robust and reliable technique for polyethylene molecular weight analysis.

Legend of Symbols

A	constant in equation 2
b	constant in equation 2
D	ordinary diffusion coefficient
D_T	thermal diffusion coefficient
F_M	mass-based fractionating power
M	polymer molecular weight
$\overline{M}_w$	weight average molecular weight
n	constant in equation 3
R	retention ratio of given component
R_s	resolution of two components
t_1	lag time prior to field decay
t_r	retention time of specified component
t^0	void time
$<v>$	mean velocity of carrier solvent
V^0	channel void volume
$\dot{V}$	volumetric flowrate
δM	difference in molecular weight between two components
ΔT	difference in temperature between hot and cold walls
ΔT_0	initial value of ΔT
λ	retention parameter
ϕ	constant in equation 3

Acknowledgments

This work was produced under Grant CHE-9102321 from the National Science Foundation. The authors thank Dr. William W. Carson of Waters Division of Millipore for providing the 150-CV instrument.

Literature Cited

1. Giddings, J. C. *Chem. Eng. News* **1988**, *66* (Oct. 10), 34-45.
2. Giddings, J. C. *Sep. Sci. Technol.* **1984**, *19*, 831-847.
3. Caldwell, K. D. *Anal. Chem.* **1988**, *60*, 959A-971A.
4. Giddings, J. C. *Anal. Chem.* **1971**, *53*, 1170A-1175A.
5. Giddings, J. C.; Kumar, V.; Williams, P. S.; Myers, M. N. In *Polymer Characterization*; Craver, C. D.; Provder, T., Eds.; Advances in Chemistry Series No. 227; American Chemical Society: Washington, D.C., 1991; Chapter 1.
6. Giddings, J. C. In *Size Exclusion Chromatography*; Hunt, B. J.; Holding, S., Eds.; Blackie and Son: Glasgow, 1989; Chapter 8.
7. Giddings, J. C. *Pure Appl. Chem.* **1979**, *51*, 1459-1471.
8. Myers, M. N.; Caldwell, K. D.; Giddings, J. C. *Sep. Sci.* **1974**, *9*, 47-70.
9. Schimpf, M. E. *J. Chromatogr.* **1990**, *517*, 405-421.
10. Kirkland, J. J.; Yau, W. W. *J. Chromatogr.* **1990**, *499*, 655-668.
11. Giddings, J. C.; Lin, G. C.; Myers, M. N. *J. Liq. Chromatogr.* **1978**, *1*, 1-20.
12. Benincasa, M. A.; Giddings, J. C. *Anal. Chem.* **1992**, *64*, 790-798.

13. Brimhall, S. L.; Myers, M. N.; Caldwell, K. D.; Giddings, J. C. *J. Polym. Sci., Polym. Lett. Ed.* **1984**, *22*, 339-345.
14. Brimhall, S. L.; Myers, M. N.; Caldwell, K. D.; Giddings, J. C. *Sep. Sci. Technol.* **1981**, *16*, 671-689.
15. Schimpf, M. E.; Giddings, J. C. *J. Polym. Sci.: Part B: Polym. Phys.* **1990**, *28*, 2673-2680.
16. Giddings, J. C.; Smith, L. K.; Myers, M. N. *Anal. Chem.* **1976**, *48*, 1587-1592.
17. Williams, P. S.; Giddings, J. C. *Anal. Chem.* **1987**, *59*, 2038-2044.
18. Freifelder, D. *Physical Biochemistry*; W. H. Freeman and Co.: San Francisco, 1982; Chapter 13.
19. Bird, R. B.; Armstrong, R. C.; Hassager, O. *Dynamics of Polymeric Liquids*; John Wiley: New York, 1977, Vol. 1; Chapter 4.
20. Gao, Y. S.; Caldwell, K. D.; Myers, M. N.; Giddings, J. C. *Macromolecules* **1985**, *18*, 1272-1277.
21. Giddings, J. C.; Smith, L. K.; Myers, M. N. *Anal. Chem.* **1975**, *47*, 2389-2394.
22. Schimpf, M. E.; Giddings, J. C. *J. Polym. Sci.: Polym. Phys. Ed.* **1989**, *27*, 1317-1332.
23. Giddings, J. C.; Caldwell, K. D. In *Physical Methods of Chemistry*; Rossiter, B. W.; Hamilton, J. F., Eds.; John Wiley: New York, 1989, Vol. 3B; pp. 867-938.
24. Hovingh, M. E.; Thompson, G. E.; Giddings, J. C. *Anal. Chem.* **1970**, *42*, 195-203.
25. Gunderson, J. J.; Caldwell, K. D.; Giddings, J. C. *Sep. Sci. Technol.* **1984**, *19*, 667-683.
26. Gunderson, J. J.; Giddings, J. C. *Macromolecules* **1986**, *19*, 2618-2621.
27. Schimpf, M. E.; Giddings, J. C. *Macromolecules* **1987**, *20*, 1561-1563.
28. Brandrup, J.; Immergut, E. H., Eds. *Polymer Handbook*, 3rd ed.; John Wiley: New York, 1989; Sect. VII.
29. Tanford, C. *Physical Chemistry of Macromolecules*; John Wiley: New York, 1965; Chapter 6.
30. Giddings, J. C.; Yoon, Y. H.; Myers, M. N. *Anal. Chem.* **1975**, *47*, 126-131.
31. Giddings, J. C.; Williams, P. S.; Beckett, R. **1987**, *59*, 28-37.
32. Williams, P. S.; Giddings, J. C. *J. Chromatogr.* **1991**, *550*, 787-797.
33. Gunderson, J. J.; Giddings, J. C. *Anal. Chim. Acta* **1986**, *189*, 1-15.
34. Schimpf, M. E.; Myers, M. N.; Giddings, J. C. *J. Appl. Polym. Sci.* **1987**, *33*, 117-135.

RECEIVED July 6, 1992

Chapter 5

Copolymer Retention in Thermal Field-Flow Fractionation

Dependence on Composition and Conformation

Martin E. Schimpf[1], Louise M. Wheeler[2], and P. F. Romeo[2]

[1]Department of Chemistry, Boise State University, Boise, ID 83725
[2]Exxon Chemical Company, Linden, NJ 07036

Earlier studies of copolymer thermal diffusion are extended here to include several new random and block copolymers of polystyrene (PS) and polyisoprene (PI). Thermal diffusion coefficients for these polymers in tetrahydrofuran and cyclohexane were obtained by thermal field-flow fractionation (ThFFF). The results confirm the dependence of thermal diffusion (and therefore ThFFF retention) on the radial distribution of monomers in the solvated macromolecule. For random copolymers and block copolymers that assume a random configuration in solution, the thermal diffusion coefficient D_T is a linear function of monomer composition. This relationship provides a basis for obtaining compositional information on such copolymers by ThFFF. For copolymers subject to radial segregation of its monomers, thermal diffusion is dominated by monomers located in the outer (free-draining) region of the solvated polymer molecule. The dependence of retention on the radial distribution of monomers provides a basis for evaluating bonding arrangements in copolymers.

The characterization of copolymers can be challenging because of the overlapping effects of composition and molecular weight distribution (MWD). Often the analyst would like to characterize both the MWD and the compositional distribution. In this case the traditional method of size exclusion chromatography (SEC) is not adequate because the separation is governed by size alone. Thus, molecular weight fractions with different compositions coelute in SEC (*1*). In contrast, ThFFF separates polymers by chemical composition as well as size, and is therefore capable of yielding both size and compositional information on copolymers.

Separation by size in ThFFF is governed by differences in the ordinary (Fick's) diffusion coefficient D of the polymer components, while separation by chemical composition results from differences in the thermal diffusion coefficient

0097–6156/93/0521–0063$06.00/0

D_T (2). Ordinary diffusion in polymer solutions is well-defined. In contrast, the phenomenon of thermal diffusion in liquids is poorly understood and not well characterized. Although equations exist relating retention to experimental parameters and transport coefficients D and D_T, values of D_T are not readily available and a model does not exist for predicting them from physicochemical parameters. Therefore, calibration curves are necessary for characterizing the MWD of polymers by ThFFF (although a single calibration point can be used provided the relationship between D and molecular weight is known). Calibration is not a problem in the analysis of homopolymers because well-characterized molecular weight standards are readily available for a variety of polymers. However, copolymer analysis is difficult because standards are not readily available and the compositional dependence of retention, contained in D_T, is not predictable. In this paper, we report on continued studies aimed at characterizing thermal diffusion, and in particular, the compositional dependence of thermal diffusion in copolymers.

Theory

One of the significant advantages of FFF separations stems from the uniform open channel geometry and the well-defined flow profile. Consequently, retention can be related directly to physicochemical parameters of the analyte material and carrier liquid. In ThFFF, the fundamental retention parameter λ is related to the temperature gradient dT/dx in the channel and the transport coefficients by

$$\lambda = \frac{D}{w\, D_T\, (dT/dx)} \tag{1}$$

where w is the channel thickness (the distance between the hot and cold walls). Parameter λ can also be related, by considering the role of the flow profile in downstream polymer transport, to the volume V_r of carrier liquid required to elute the polymer zone. For parabolic flow (3)

$$V^o/V_r = 6\lambda[\coth(1/2\lambda) - 2\lambda] \tag{2}$$

where V^o is the geometric (void) volume of the channel. In ThFFF equation 2 must be corrected somewhat to account for the departure from parabolic flow induced by the temperature gradient and attendant viscosity changes in the channel (4). By using equation 1 and the corrected from of equation 2, accurate values of D/D_T can be calculated from the measured retention volume V_r.

As V_r increases, the bracketed term on the right side of equation 2 approaches unity. When $V_r > 3\ V^o$, V^o/V_r can be approximated by $V^o/V_r = 6\lambda$. In this case, combining equations 1 and 2 and approximating dT/dx as $\Delta T/w$, where ΔT is the temperature difference between the hot and cold walls, yields

$$V_r = \left(\frac{V^o\, \Delta T}{6}\right)\ \left(\frac{D}{D_T}\right) \tag{3}$$

Equation 3 shows that for $V_r > 3\ V^o$ and constant field strength (ΔT), retention is a linear function of ratio D/D_T.

The diffusion coefficient for dilute solutions is governed by the Stokes-Einstein equation, which can be related to the intrinsic viscosity $[\eta]$ by (5)

$$D = \left(\frac{kT}{6\pi\eta_0}\right) \left(\frac{10\pi N}{3M[\eta]}\right) \tag{4}$$

where η_0 is the solvent viscosity, M is the polymer molecular weight, and N is Avagadro's number. In contrast to D, the physicochemical parameters governing thermal diffusion in liquids are unknown. This is unfortunate because if D_T values were available, ThFFF could be used with a viscosity detector to obtain MWD information directly, without calibration. The lack of information on thermal diffusion also limits access to compositional information on copolymers, available from the dependence of ThFFF retention on thermal diffusion. We note, however, that detectors selective to composition (such as an infrared spectrometer) can be used to obtain average composition as a function of ThFFF retention time.

Background

Recent advances in ThFFF instrumentation and methodology have made the advantages to understanding thermal diffusion more acute. As a result, a systematic study was begun on the thermal diffusion of polymer solutions (6-8). The success of these studies, which we summarize next, has been aided by the ability of ThFFF to produce accurate values of thermal diffusion parameters using sub-milligram quantities of polymer.

In 1987 Schimpf and Giddings (6) demonstrated that D_T is independent of both molecular weight and branching configuration. Next, D_T values were obtained for 17 polymer-solvent systems (7). The results were used to correlate D_T with several polymer and solvent parameters, including the thermal conductivities of the polymer and solvent, the polymer density, and the viscosity and viscous activation energy of the carrier liquid. Independent reports were also made correlating polymer thermal diffusion with the solvating power of the solvent (9,10).

More recently, the thermal diffusion of several copolymers in toluene was characterized (8) and related to that of corresponding homopolymers. The intention was to look for additional effects influencing the thermal diffusion of copolymers. For example, it was unclear whether thermal diffusion is affected equally by all the monomer units, or if the effect is dominated by monomers located in specific regions, such as the outer free-draining region of the solvation sphere. Block copolymers are useful probes in addressing this issue because their monomers are subject to radial segregation. For example, segregation of blocks containing different monomers a and b can be built into the primary structure of a copolymer during synthesis, particularly in star-shaped copolymers prepared by joining the ends of linear diblock arms. In this case, the monomers located proximal to the junction of the arms are physically anchored to the inner region of the solvated molecule. Segregation is also induced in the secondary structure of block copolymers dissolved in a liquid that is a better solvent for one of the monomer types (11). In such a liquid, subsequently referred to as a selective solvent, the more soluble polymer segments migrate to the outer regions of the solvated molecule, surrounding a more condensed core containing less soluble segments. The presence of solvent-induced segregation has been supported by fluorescence-emission probe techniques and low-angle light scatering methods (12).

In studying the effect of monomer location on thermal diffusion (8), it was found that in random copolymers, D_T values apparently assume the weighted average of the corresponding homopolymer values, where the weighting factors are the mole-fractions of each monomer type in the copolymer. This is in marked contrast with block copolymers subject to radial segregation of their monomers, where thermal diffusion appears to be dominated by monomers located in the outer (free-draining) region of the solvated polymer molecule. In the work reported here, values of D_T were obtained for PS-PI linear block copolymers of varying composition in cyclohexane and THF. Cyclohexane is selective for PI, since it is a good solvent for PI and a theta-solvent for PS. In contrast, THF is an equally good (non-selective) solvent for PS and PI. In cyclohexane, PI segments are expected to segregate to the exterior of the solvated molecule; in THF, monomer segregation is not expected. Integrated into this report are the results in toluene already published by Schimpf and Giddings (12), so that a complete and current summary of copolymer retention in ThFFF is presented.

Experimental Outline

Polymer Samples. Detailed information on the polymers utilized in these studies is contained in Table I. All samples contain narrow molecular weight distributions (polydispersity less than 1.1). The copolymers can be divided into groups that differ in one major aspect. The differences include monomer ratios in both random and block copolymers, block arrangement in block copolymer pairs having both linear- and star-shaped configurations, and arm number and arm molecular weight in star-shaped copolymers. By examining the influence (or lack of influence) of these distinctive features on D_T values, the phenomenon of thermal diffusion in copolymers is characterized. Samples were typically dissolved in the carrier liquid at a concentration of 1 mg/ml.

Instrumentation. The ThFFF system used in this work is similar to the model T100 polymer fractionator from FFFractionation Inc. (Salt Lake City, UT), with a channel length of 46 cm (tip-to-tip), a width of 2.0 cm, and a thickness of 127 μm. The SEC column used to measure polymer diffusion coefficients is a commercial Ultrastyragel column from Waters Chromatography Div., Millipore Corp. (Milford, MA). Detection was accomplished with a refractive index monitor for the toluene work; otherwise, a fixed-wavelength (254 nm) UV detector was used.

Determination of Diffusion Coefficients. In order to determine D_T values from ThFFF retention data, accurate values for the polymer diffusion coefficient D must be available. Values of D in THF and toluene were obtained by calibrating an SEC column as ln D versus retention volume using a series of polystyrene standards whose D values were obtained from light scattering data (7). A separate calibration curve was prepared for each solvent because the pore size of the column-packing material changes with solvent. Values of D in cyclohexane were obtained directly by dynamic light scattering at Exxon's research laboratories in Clinton, New Jersey.

Table I. Summary of Polymers Examined

Polymer No.	MW (daltons)	Composition[a] (mol-%)	shape	bonding arrangement[b]
1.	c	100 PS	linear	----
2.	c	100 PI	linear	----
3.	c	100 PM	linear	----
4.	34,000	72 PS, 28 PM	linear	random
5.	unknown	57 PS, 78 PM	linear	random
6.	600,000	22 PS, 78 PI	star-18[d]	PI inside
7.	600,000	22 PS, 78 PI	star-18	PS inside
8.	300,000	57 PS, 43 PI	linear	triblock, PI inside
9.	300,000	57 PS, 43 PI	linear	diblock
10.	76,000	22 PS, 78 PI	star-8	PI inside
11.	113,000	22 PS, 78 PI	star-12	PI inside
12.	833,000	22 PS, 78 PI	star-12	PI inside
13.	28,000	77 PS, 23 PI	linear	diblock
14.	39,000	43 PS, 57 PI	linear	diblock
15.	40,000	16 PS, 84 PI	linear	diblock
16.	38,000	51 PS, 49 PI	linear	random

a. PS = polystyrene, PM = polymethylmethacrylate, PI = polyisoprene
b. all arms of star-shaped polymers are linear diblocks
c. includes a variety of polymer standards of different molecular weights
d. number of arms

Results and Discussion

The D_T values obtained in these studies are summarized in Table II. The standard error in these values is 6%. Figure 1 illustrates the approximate linear dependence of D_T on the relative content of methylmethacrylate in PS-PMMA random copolymers dissolved in toluene, using homopolymer D_T values as endpoints. Monomer segregation is not possible in random copolymers because they do not contain long segments consisting of one monomer type.   In this case the data indicates that copolymer D_T values can be described as a weighted average of the D_T values of the homopolymer constituents, where the weighting factors are the mole fractions X_a and X_b of the constituent monomers.   This relationship can be expressed as

$$D_T(ab) = X_a D_T(a) + X_b D_T(b) \qquad (5)$$

where $D_T(ab)$, $D_T(a)$, and $D_T(b)$ are the thermal diffusion coefficients of ab random copolymer, a homopolymer, and b homopolymer, respectively.   Because of the limited amount of data represented by equation 5, other relationships cannot be ruled out, including a sigmoidal dependence of $D_T(ab)$ on $D_T(a)$ and $D_T(b)$ and relationships that utilize different weighting factors such as vol-%.

A similar plot of D_T values versus composition for block copolymers of PS and PI in toluene shows no correlation between D_T and monomer composition, as illustrated in Figure 2. In every copolymer except no. 7, there is a deviation in D_T from the linear dependence observed with random copolymers toward the D_T value of PS homopolymer. We now examine these D_T values individually to see if this departure from linearity can be consistently explained by monomer segregation.

The "ideal" linear relationship between D_T and styrene content based on the homopolymer D_T values (expressed in cm^2/s-K) can be written as

$$D'_T \ (X \ 10^7) = 0.69 + 0.34 \ X_S \tag{6}$$

Table II. Summary of D and D_T values for Copolymers and Component Homopolymers

Polymer No.	Solvent	D X 10^7 (cm^2/s)	D_T X 10^7 (cm^2/s-K)
1. PS	toluene	a	1.03
	cyclohexane	a	0.44
	THF	a	0.94
2. PI	toluene	a	0.69
	cyclohexane	3.40	0.08
	THF	a	0.51[b]
3. PM	toluene	a	1.63
4. PS-PM	toluene	5.71	1.12
5. PS-PM	toluene	5.77	1.33
6. PS-PI	toluene	2.74	0.93
7. PS-PI	toluene	3.09	0.74
8. PS-PI	toluene	2.87	1.08
9. PS-PI	toluene	2.95	0.97
10. PS-PI	toluene	4.80	0.83
11. PS-PI	toluene	4.42	0.84
12. PS-PI	toluene	2.84	0.84
13. PS-PI	cyclohexane	1.89	0.12
	THF	11.8	0.87
14. PS-PI	cyclohexane	1.53	0.15
	THF	9.69	0.69
15. PS-PI	cyclohexane	0.52	0.05
	THF	9.52	0.58
16. PS-PI	cyclohexane	2.84	0.29
	THF	9.79	0.75

a. includes a variety of polymer standards of different molecular weight
b. data obtained from reference 1

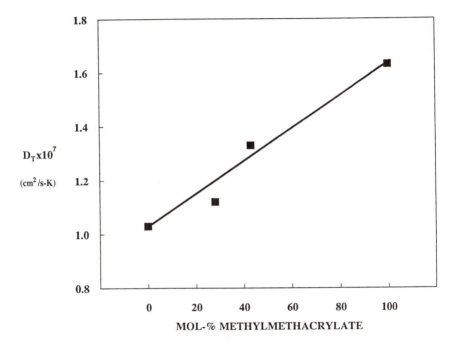

Figure 1. Plot of D_T versus mol-% methylmethacrylate in random copolymers of styrene and methylmethacrylate (4 and 5 in Table I) in toluene. (Adapted from ref. 7. Copyright 1989 John Wiley & Sons)

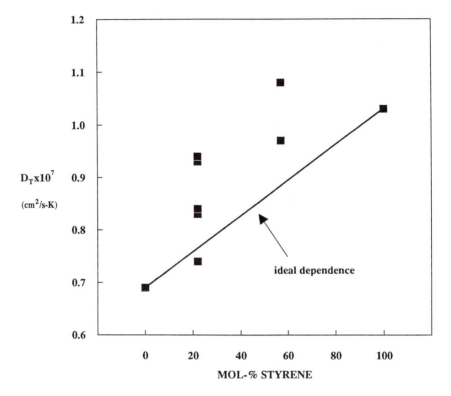

Figure 2. Plot of D_T versus mol-% styrene for block copolymers of styrene and isoprene (6 through 12 in Table I) in toluene
(Adapted from ref. 7. Copyright 1989 John Wiley & Sons)

where X_S is the mole-fraction of styrene monomer. The percent deviation ψ of measured D_T values from those predicted by equation 6 is given by

$$\psi = \frac{D_T - D'_T}{D'_T} \times 100 \tag{7}$$

Values of D_T, D'_T, and ψ for block copolymers 6-12 are shown in Table III.

As explained earlier, star copolymers that consist of di-block arms are subject to segregation due to bonding constraints and solubility effects. The degree of segregation in a given copolymer depends on the length, number, and arrangement of the homogeneous segments, and the magnitude of the differences in the solubilities of the constituent homopolymers. If we consider a linear polymer as a star polymer with only two arms, we expect that linear block copolymers are also subject to radial segregation due to bonding constraints, although to a lesser extent. The tendency for PS-PI copolymers to segregate in toluene due to solvent effects is ambiguous because the relative solubilities of PS and PI in toluene are not available. Considering the similarity of toluene to the PS repeat unit, we expect toluene to be a selective solvent for PS. This is supported by data on the solubility parameters of PS, PI, and toluene (8). However, Mark-Houwink exponents (relating intrinsic viscosity to molecular weight) are similar for PS and PI in toluene (13), indicating that toluene is an equally good solvent for both homopolymers. With these considerations, we now examine ψ values in relation to both segregation effects.

Copolymers 6 and 7 are star-shaped copolymers containing 18 arms, each arm being a diblock with 22 mol-% styrene. The only difference between these copolymers is the location of the styrene block relative to the center of the molecule. In copolymer 6 the diblock arms are joined at the PI end, confining a portion of the PI segments to the center of the solvated molecule. The measured D_T value deviates significantly ($\psi = 23\%$) from the "ideal" value predicted by equation 6. Thus, the

Table III
Percent Deviation From Equation 6 of Measured D_T Values
for PS-PI Copolymers in Toluene

Polymer (no.)	$D_T \times 10^7$ (cm^2/s-K)	$D'_T \times 10^7$ (cm^2/s-K)	ψ (%)
6	0.93	0.76	22
7	0.74	0.76	-3
8	1.08	0.88	23
9	0.97	0.88	10
10	0.83	0.76	9
11	0.84	0.76	11
12	0.94	0.76	24

D_T value of 0.93 x 10^{-3} cm^2/s-K for copolymer 6 is much closer to PS homopolymer (1.03) than PI homopolymer (0.69) even though the polymer contains only 22 mol-% styrene. In contrast, the arms of copolymer 7 are joined at the PS end, confining part of the PS segments to the inner region of the star configuration, restricting any solvent-based segregation that may be present. This restriction is severe enough to result in a slight negative deviation in D_T (ψ = -3%). The significantly different values of D_T for copolymers 6 and 7 indicate that thermal diffusion is influenced by the radial distribution of monomers in the solvated molecule, with outer monomers having a greater influence.

The smaller deviation in D_T from equation 6 for copolymer 7 compared to copolymer 6 is consistent with the presence of solvent-induced segregation. Thus, in copolymer 6, both segregation effects may act in concert, while in copolymer 7, solvent-induced segregation may nearly cancel the effect of bonding constraints. We repeat, however, that the presence of solvent-induced segregation is uncertain. Therefore, while the radial distribution is clearly a factor in thermal diffusion, other unknown factors may responsible for the significant difference in the absolute values of ψ for copolymers 6 and 7.

More evidence of the dominant influence of the outer region comes from examining copolymers 8 and 9. These are linear block copolymers containing 43 mol-% isoprene. In order to compare the effect of bonding constraints in these copolymers to the same effects in copolymers 6 and 7, we consider copolymers 8 and 9 to be star polymers with only two arms. In copolymer 8, two equivalent diblock arms are joined at the isoprene ends, confining part of the PI segments to the center of the solvated molecule. In copolymer 9, a homogeneous PS arm is joined to the PS end of a diblock arm, and a fraction of the PS monomers are confined to the core of the solvated molecule. The D_T value of copolymer 8 (1.08 x 1010^{-7} cm^2/s-K) is significantly greater than that of copolymer 9 (0.97 x 10^{-7} cm^2/s-K), lending further support to the theory that a given monomers role in thermal diffusion becomes more significant as the radial distance of the monomer from the core of the solvated polymer molecule increases.

The positive deviation in D_T from equation 6 (ψ = 10%) is unexpected because any segregation in the primary structure should produce a negative deviation. Still, the difference in the D_T values of copolymers 8 and 9 (and copolymers 6 and 7) indicate that outer monomers are more dominant in thermal diffusion. The greater influence of PS than expected from considering only bond-induced segregation is therefore due to solvent-induced segregation or other phenomena. The "other" phenomena, whether solvent induced or not, are clearly more influential in the linear copolymers. This is expected since bond-induced segregation is greater in the star-shaped copolymers.

The effect of arm length and number on thermal diffusion in copolymers is unclear. In PS homopolymers, D_T is independent of arm number and length (6). In copolymers, monomer segregation is presumably enhanced by a larger number of shorter arms. Therefore, we expect D_T to be higher for copolymer 11 compared with copolymer 10; this is the case. However, our expectation that D_T for copolymer 11 should be higher than that for copolymer 12 is not realized. Moreover, copolymer 12 has a D_T value that is virtually identical to that of copolymer 6, even though the latter has larger number of shorter arms.   The anomolous behavior of copolymer 12 requires further study.

To explore the effect of solvent-induced segregation more conclusively, we examined several diblock copolymers of PS and PI in THF and cyclohexane. Like toluene, THF is expected to be an equally good solvent for PS and PI, based on Mark-Houwink exponents. Cyclohexane, on the other hand, is a selective solvent for PI, being a poor theta solvent for PS.   Therefore, we expect solvent-induced segregation of PI segments to the exterior of PS-PI block copolymers dissolved in cyclohexane.

In THF, we find that thermal diffusion behaves in a similar manner as random copolymers.   Thus, D_T values are a linear function of the mol-% composition of styrene, as illustrated in Figure 3. This is in marked contrast with cyclohexane, where the linear dependence vanishes, as illustrated in Figure 4. We observe that the D_T values for block copolymers 13-15 in cyclohexane are closer to that of PI, which is better solvated by the carrier liquid.   Finally, in random copolymer 16, where monomer segregation is precluded by the absence of long segments that contain only one monomer type, the measured D_T value falls in line with the homopolymer values, indicating that thermal diffusion is equally influenced by both monomers types. These results imply that solvent-induced segregation is a factor in thermal diffusion in a manner consistent with an increasing influence from monomers located in the free-draining region of the molecule.

Conclusions

Thermal diffusion of PS-PI copolymers in dilute solutions is influenced by the radial distribution of monomer segments.   In random copolymers, where large homogeneous segments do not exist, the relative influence of each monomer type corresponds to its mole-fraction in the polymer molecule. In this case, equations can be established relating thermal diffusion to monomer content. These equations can be used to determine unknown compositions once the D_T values of constituents homopolymers are obtained. Thermal diffusion (and therefore retention) of two-component block copolymers share the same relationship to monomer content as random copolymers only when the monomers are randomly distributed in the molecules solvation sphere.   Two requirements exist for a random distribution. First, a carrier liquid must be used that is an equally good solvent for both monomers. Second, the ability of the monomers to distribute randomly about the solvation sphere cannot be significantly restricted by bonding constraints.

While complete randomization is impossible in any block configuration, such restrictions are apparently not significant for diblocks in non-selective solvents, so that thermal diffusion follows the same linear dependence on monomer composition as in random copolymers. Branched structures, on the other hand, contain more severe bonding constraints to randomization, but it is not clear how differences in branching configuration affect thermal diffusion.   Segregation of homogeneous segments in block copolymers dissolved in a selective solvent provides a basis for using ThFFF to study bonding arrangements in linear copolymers, such as the degree of "blockiness".   The impact on thermal diffusion of segregation in primary structure gives ThFFF potential to characterize bonding in branched structures, for

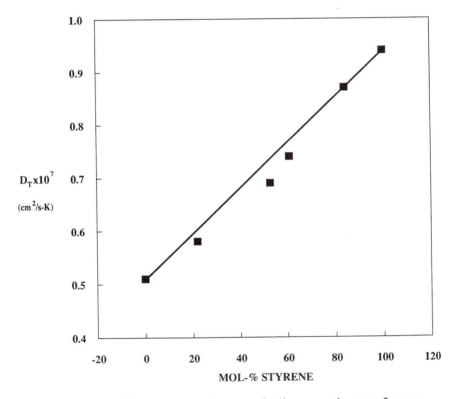

Figure 3. Plot of D_T versus mol-% styrene for linear copolymers of styrene and isoprene (13 through 16 in Table I) in THF.

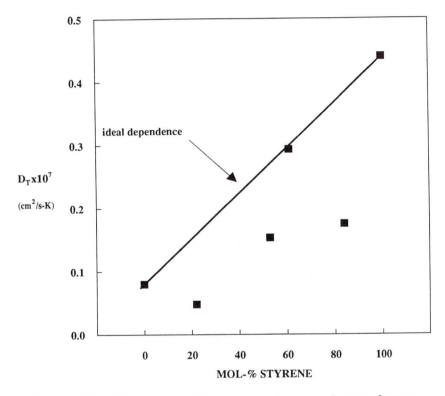

Figure 4. Plot of D_T versus mol-% styrene for linear copolymers of styrene and isoprene (13 through 16 in Table I) in cyclohexane.

example the degree of grafting in graft copolymers, although more studies are needed to determine the extent of this capability.

Acknowledgements

The authors gratefully acknowledge the gift of copolymer samples by Dr. J. Dias, Mr. J. Olkusz, Dr. S. Mori, Dr. E. Thomas, and Dr. J. Shen. The generous support of the Field-Flow Fractionation Research Center (University of Utah, Salt Lake City, Utah), where much of this work was performed, is also acknowledged. This work was supported in part by Grant CHE-9102321 from the National Science Foundation. M.E.S. is also pleased to acknowledge grants from the Idaho State Board of Education (93-030) and Research Corporation (C-3272) for support of this research.

Literature Cited

1. Gunderson, J. J.; Giddings, J. C. *Macromol.* **1986**, *19,* 2618-2621.
2. Schimpf, M. E. *J. Chromatogr.* **1990**, *517*, 405-421.
3. Giddings, J. C.; Caldwell, K. D.; Myers, M. N. *Macromol.* **1976**, *9,* 106.
4. Gunderson, J. J.; Caldwell, K. D.; Giddings, J. C. *Sep. Sci. Technol.* **1984**, *19,* 667.
5. Rudin, A.; Johnston, H.K. *J. Polymer Sci., Part B* **1971**, *9,* 55.
6. Schimpf, M. E.; Giddings, J. C. *Macromol.* **1987**, *20,* 1561-1563.
7. Schimpf, M. E.; Giddings, J. C. *J. Polym. Sci., Polym. Phys. Ed.* **1989**, *27,* 1317-1332.
8. Schimpf, M. E.; Giddings, J. C. *J. Polym. Sci., Polym. Phys. Ed.* **1990**, *28,* 2673-2680.
9. Schimpf, M. E.; Giddings, J. C. presented at the *192n d National ACS Meeting,* **1986,** Anaheim, CA.
10. Kirkland, J. J. presented at the *First International Symposium on Field-Flow Fractionation* **1989,** Park City, Utah.
11. Sadron, C. *Angew. Chem. Intern. Ed.* **1983**, *2,* 248.
12. Watanabe, A.; Matsuda, M. *Macromol.* **1985**, *18,* 273.
13. Brandup, J.; Immergut, E. H., Eds., in *Polymer Handbook*, Wiley, New York, **1964**.

RECEIVED April 20, 1992

Chapter 6

Gel-Content Determination of Polymers Using Thermal Field-Flow Fractionation

Seungho Lee

Analytical Research, 3M Corporate Research Laboratories, St. Paul, MN 55144

The object of this study is to investigate the applicability of thermal field-flow fractionation (ThFFF) for characterization of gel-containing polymers. ThFFF was used to monitor the effect of electron beam (EB) treatment on high molecular weight poly methyl methacrylate (PMMA) samples. A trend of polymer degradation was clearly observed; as the dose of electron beam increases, a low molecular weight shoulder grows and the average molecular weight decreases. No sign of polymer cross-linking was observed within the range of the electron beam intensity of up to 10 Mrad. ThFFF was also used to determine the difference between two acrylate elastomers which were manufactured by the same procedure but show different mechanical properties. A study using SEC, SEC-Viscometry, and SEC-Light scattering photometry found no significant differences in molecular weight and in degree of branching. ThFFF channel is open (not packed) and a sample can be injected without filtration. Gel content of a sample was determined by subtracting the peak area of filtered sample from that of unfiltered one. ThFFF result shows significant difference in gel content between two samples, 17 to 7 %.

Thermal field-flow fractionation (ThFFF) is a separation technique for a wide range of synthetic polymeric materials (*1-5*). Separation is carried out by applying a temperature gradient across a thin, ribbon-like flow channel (Figure 1A). Under the temperature gradient, the solute molecules are driven toward the cold wall by the thermal diffusion process. Accumulation of the solute by thermal diffusion is counteracted by ordinary diffusion and a steady state distribution of the solute is eventually established near the cold wall. Once all the solutes reach their steady state, the channel flow is initiated and the solutes are carried down the channel by the flow. Solutes driven closer to the cold wall experience a lower flow rate region

0097–6156/93/0521–0077$06.00/0

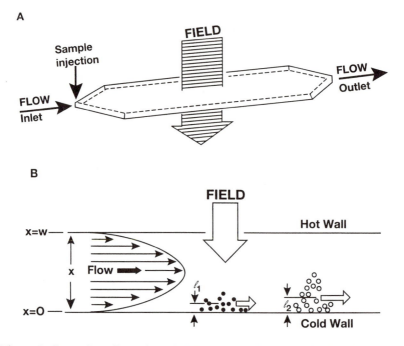

Figure 1. General configuration of FFF channel (A), and in-depth side view of ThFFF channel (B).

of the parabolic flow profile and their average down stream migration velocity is slower than those with greater layer thicknesses. For the polymers with the same chemical structure, the ordinary diffusion coefficient decreases as the molecular weight increases, while the thermal diffusion coefficient remains approximately constant. High molecular weight components are thus forced closer to the cold wall and elute after low molecular weight components (Figure 1B).

Both ThFFF and size exclusion chromatography (SEC) can be used for determination of molecular weight and molecular weight distribution of polymers. The ThFFF channel is open and provides advantages over SEC for the characterization of high molecular weight polymers or gel-containing polymers. First, because the ThFFF channel is unobstructed, the channel flow is well-characterized and basic parameters such as retention and zone broadening are easily predicted by theory. Theoretical predictability is essential for fast system optimization. Second, there is virtually no extentional shear in ThFFF due to the absence of packing material. Thus shear degradation or shear-induced structural changes of polymers are less likely than in SEC. SEC suffers from degradation of polymer chains due to high-shear flow of the packed bed (*6-9*). Shear degradation becomes more serious as the polymer molecular weight increases. ThFFF is thus advantageous for the analysis of very high molecular weight polymers. Third, sample filtration is not required in ThFFF. Sample solution is usually filtered in SEC to prevent column damage such as blockage and contamination, etc. Sample filtration is highly undesirable especially when the sample contains components that are removed by filtration (such as gel particles). SEC, SEC-Viscometry, and Light scattering photometry are widely used for polymer solution characterization (*10*) but sample filtration is usually required. Fourth, ThFFF has no packing material with which the sample components can interact. Gel particles that are not completely dissolved in solution tend to adsorb on the packing material of SEC columns. Even some soluble polymers can interact with the packing material and adsorb.

Assuming negligible sample loss in the ThFFF channel, peak areas obtained from a concentration detector (such as UV-VIS or refractive index detector) are directly related to the mass of the sample injected. The gel content of a polymer sample can be quantitatively determined from the difference in peak area between the filtered and unfiltered solution of the same sample. In this paper, the applicability of ThFFF to the characterization of gel-containing polymers will be investigated.

Theory

There have been numerous publications describing the basic separation mechanism of ThFFF. Briefly covered here is how ThFFF retention data is related to the molecular weight of the sample. As mentioned previously, the extent of retention of a solute in ThFFF depends on the thickness of the equilibrium layer formed as a result of two counteracting diffusion processes (thermal and ordinary diffusion). The thicker the layer is, the faster the solutes elute. The reduced thickness, λ is defined as the layer thickness divided by the channel thickness, w, and is given (11) by

$$\lambda = \frac{D}{D_T} \left[\frac{1}{w(dT/dx)} \right] \qquad (1)$$

where D is ordinary diffusion coefficient, D_T is the thermal diffusion coefficient, x is the distance from the cold wall (see Figure 1B) and dT/dx is the temperature gradient applied across the channel. It can be seen that the ratio, D/D_T is the determining factor for the solute layer thickness (and thus the solute retention). For a given polymer/solvent combination, D_T is approximately independent of polymer molecular weight while D decreases with increasing molecular weight. It can be assumed in ThFFF (6) that

$$\frac{D}{D_T} = \phi M^{-n} \qquad (2)$$

and

$$w\left(\frac{dT}{dx}\right) = \Delta T \qquad (3)$$

where ϕ is a constant for a given polymer/solvent combination at a fixed cold wall temperature, n is a constant that generally takes a value of approximately 0.6 and ΔT is the temperature difference between the hot and cold walls. Substituting equations 2 and 3 into equation 1 gives

$$\lambda = \frac{\phi M^{-n}}{\Delta T} \qquad (4)$$

Retention ratio R is defined as the ratio of the zone migration velocity to the average carrier velocity. Assuming it is an isoviscous condition, R is given (12) by

$$R = 6\,\lambda \left[\coth\left(\frac{1}{2\lambda}\right) - 2\lambda \right] \qquad (5)$$

which simplifies to

$$R = 6\lambda \qquad (6)$$

for the case of high retention (small value of λ). Experimentally the retention ratio of a component can be determined (6) by

$$R = \frac{V^o}{V_r} \qquad (7)$$

where V^o is channel volume and V_r is the observed elution volume of the component. Combining equations 4, 6 and 7 gives

$$\frac{V^o}{V_r} = \frac{6\phi M^{-n}}{\Delta T} \qquad (8)$$

Using equation 8, the solute molecular weight, M can be determined from the experimental data, V_r, if ϕ and n are available.

The constants ϕ and n, are usually determined by running a series (or a mixture) of standards. The logarithmic form of equation 2 is

$$\log\left(\frac{D}{D_T}\right) = \log \phi - n \log M \qquad (9)$$

For a standard of known molecular weight M, D/D_T can be calculated from its retention ratio using equations 1, 5 and 7. In the plot of $\log (D/D_T)$ vs. $\log$ M, ϕ and n are determined from the intercept and the slope respectively.

Experimental

The ThFFF system used in this study is a Polymer Fractionator model T100 from FFFractionation, Inc. (Salt Lake City, Utah). The channel has dimensions of 0.0127 cm in thickness, 1.9 cm in breadth, and 45.6 cm tip to tip length. The channel volume, V^o is 1.05 mL. The channel effluent is monitored by a HP 1037A refractive index detector (Hewlett Packard, Palo Alto, CA). The system is equipped with a Rheodyne 20 μL loop sample injector.

The SEC column is a 500 x 10 mm Permagel column (Column Resolution, Inc., San Jose, CA). It is a mixed bed column with the particle diameter of 10 μm. Three detectors are used for SEC. These include an HP 1037A Refractive Index detector(RID), Model 100 differential viscometer (Viscotek, Porter, Texas) and a KMX-6 Low Angle Laser Light Scattering Photometer (Milton Roy, Arlington, IL).

Polymer molecular weight standards were narrow polystyrenes obtained from Polymer Laboratories, Ltd (Amherst, MA). High molecular weight polymethyl methacrylate (PMMA) samples were commercial PERSPEX CQ/UV (Imperial Chemical Industries, Wilmington, DE) treated with an electron beam of various intensities using a CB 300/30/380 (Energy Science Inc., Woburn, MA). Acrylate elastomers were synthesized within 3M laboratories. They have acid groups that may interact with SEC column packing material (*13*). For SEC experiments, the acrylate elastomers were methylated using in-house synthesized diazomethane to prevent sample adsorption. Tetrahydrofuran (HPLC reagent, J.T. Baker Inc., Phillipsburg, NJ) was used as a carrier fluid for all experiments. The flow rate was 1.0 mL/min for SEC and 0.2 mL/min for ThFFF experiments. The polymer solutions in THF had concentrations of approximately 0.1 - 0.2 % (w/v). For sample filtration, a 0.2 μm PTFE disposable filter (nonsterile, 25 mm disc) was used.

Result and discussion

In a power programming mode of ThFFF, the temperature gradient, ΔT, is continuously decreased during a run according to a power function. The power function is specified by parameters t_a, t_1 and p which are the time constant, predecay time, and decay power respectively (14, 15). According to theory, uniform resolution over the full molecular weight range of the sample can be achieved when $t_a = -2 t_1$. A number of programmed runs were made with mixtures of various combination of polystyrene standards. Figure 2 shows the plot of log (D/D_T) vs. log M obtained from a mixture of four narrow polystyrene standards whose nominal molecular weights are 4.7×10^4, 4.9×10^5, 2.05×10^6 and 9.35×10^6. The power programming parameters were: initial $\Delta T = 60$ °C, $t_1 = 4$ min, $t_a = -8$ min and flow rate = 0.2 mL/min. The circles represent the log (D/D_T) values calculated for each standard as described earlier and the straight line is the result of linear regression of those four data points. The plot shows an excellent linearity for entire molecular weight range of the mixture. The calculated values of the constants are $\phi = 1.29 \times 10^4$ and n = 0.647. It was observed that the values of both ϕ and n generally increase as t_1 increases with other programming parameters kept constant.

ThFFF was used to investigate degradation of high molecular weight PMMA exposed to an electron beam (EB). Figure 3 shows fractograms of EB-treated PMMAs obtained at the same condition as in Figure 2. Figure 3A is the fractogram of untreated PMMA. Doses of the electron beam were gradually increased from 1 Mrad for Figure 3B to 10 Mrad for Figure 3E. Number and weight average molecular weights determined for each fractogram are shown in Table I.

A clear trend of degradation is observed with increasing dose of electron beam: as the dose of electron beam increases, a low molecular weight shoulder grows and the elution profile becomes bimodal. The weight average molecular weight decreases from 6 million to approximately 3 million. It is known that electron beam treatment on polymers could result in cross-linking of polymer chains and thus cause an increase in the average molecular weight of the polymers. No sign of chain cross-linking was observed within the range of electron beam of up to 10 Mrad.

ThFFF was also used to determine the difference between two acrylate elastomers manufactured by the same procedure, but showing different mechanical behaviors. They are arbitrarily labeled 'normal' and 'abnormal' to distinguish one from another. When SEC was used, no significant difference was found between

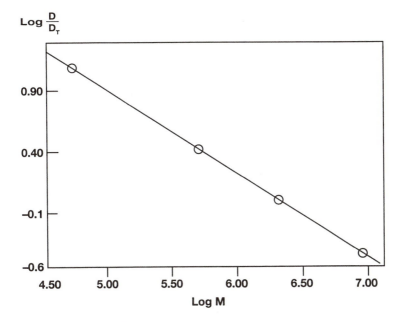

Figure 2. Plot of log (D/D$_T$) vs. log M obtained from a power programmed ThFFF run of a mixture of four polystyrene standards (nominal molecular weight = 4.7 x 10^4, 4.9 x 10^5, 2.05 x 10^6 and 9.35 x 10^6).

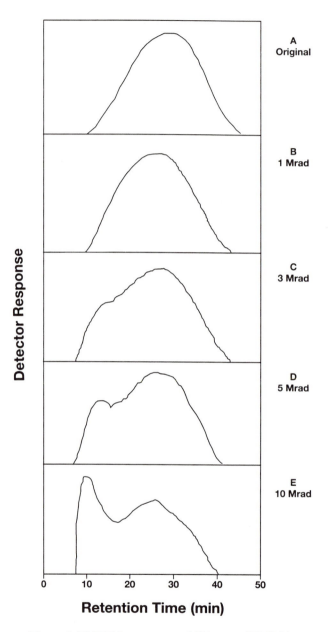

Figure 3. ThFFF fractograms of EB-treated PMMAs.

two samples in molecular weight and degree of branching. Table II shows the molecular weight data obtained from SEC-RID, SEC-viscometry and SEC-low angle laser light scattering photometry (SEC-LS).

Table I. Molecular weight of EB-treated PMMAs determined by ThFFF

EB dose (Mrad)	M_n^a	M_w^b
Untreated	2.29×10^6	6.04×10^6
1	1.49×10^6	4.61×10^6
3	8.67×10^5	4.21×10^6
5	4.91×10^5	3.71×10^6
10	3.24×10^5	3.02×10^6

[a] number average molecular weight
[b] weight average molecular weight

Table II. Molecular Weight of Acrylate elastomers determined by various SEC-related techniques

Sample	SEC/RID		SEC/Viscometry		SEC/LS	
	M_n	M_w	M_n	M_w	M_n	M_w
Normal	2.58×10^5	9.20×10^5	1.83×10^5	1.29×10^6	4.99×10^5	8.30×10^5
Abnormal	1.92×10^5	9.40×10^5	1.99×10^5	1.31×10^6	5.02×10^5	9.72×10^5

SEC and Light Scattering can be used together to determine the degree of branching of a polymer. SEC separates polymer molecules according to their sizes (or hydrodynamic volumes). For a given molecular weight, the dimension of a polymer molecule decreases as the degree of branching increases. Thus the ratio of the actual molecular weight (which can be determined by light scattering photometry) to the apparent molecular weight determined by SEC is a measure of branching of the polymer. The higher the ratio, the more chain branching there is in the polymer. The ratio is 0.902 for normal and 1.03 for abnormal sample.

Figure 4 shows the molecular weight distributions determined by ThFFF for the normal (top) and abnormal (bottom) acrylate elastomers. The programming parameters were: initial $\Delta T = 90$ °C, $t_1 = 3$ min, $t_a = -6$ min and flow rate = 0.2

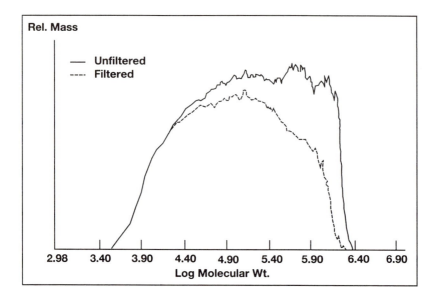

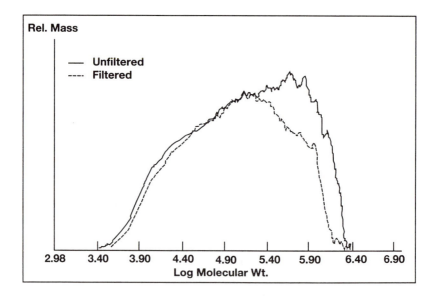

Figure 4. Molecular weight distribution of normal (top) and abnormal (bottom) acrylate elastomers.

mL/min. For each sample, the molecular weight distribution of the unfiltered solution is shown as a solid line and that of filtered solution as a dotted line. In both elastomers, some of the high molecular weight components were removed by filtration. Assuming the excess detector response obtained from the unfiltered solution is only due to the gel particles, the gel content of a sample was determined by dividing the excess peak area by the peak area of the unfiltered solution. The average molecular weight and gel content determined by ThFFF are shown in Table III. Significant difference in gel content was observed between normal (17.4 % gel) and abnormal (7.0 % gel) elastomers.

Table III. Molecular Weights and Gel Contents of Acrylate elastomers determined by ThFFF

Sample	M_n	M_w	Gel Content (%)
Normal	4.06×10^5	1.03×10^6	17.4
Normal(filtered)	2.43×10^5	6.59×10^5	-
Abnormal	3.91×10^5	9.66×10^5	7.0
Abnormal(filtered)	2.54×10^5	6.49×10^5	-

Conclusion

The results of this preliminary study show that ThFFF can be used for the characterization of gel-containing polymers. The method for determining gel content is based on a few assumptions: (1) all the gel particles are removed by filtration and the excess peak area obtained from the unfiltered sample is only due to the gels, (2) the detector signal is proportional to the polymer concentration throughout the whole molecular weight range of the sample including the gels, (3) there is no polymer degradation due to filtration, etc. Because of these assumptions, the gel content data presented in this paper may contain inaccuracies, and the application of this method should be limited to comparative studies only. This study provides a foundation for further ThFFF studies on gel-containing polymers, including gel characterization. The use of an evaporative light scattetring detector(instead of RI or UV-VIS) may result in more accurate gel content data in this method.

Acknowledgments

The author wishes to thank Dr. Albert Martin of 3M company for the acrylate elastomer samples.

Literature Cited

(1) Thompson, G. H.; Myers, M. N.; Giddings, J. C. *Anal. Chem.* **1969**, *41*, 1219.

(2) Hovingh, M. E.; Thompson, G. H.; Giddings, J. C. *Anal. Chem.* **1970**, *42*,195.

(3) Gunderson, J. J.; Giddings, J. C. *Macromolecules* **1986**, *19*, 2618.

(4) Kirkland, J. J.; Yau, W. W. *Macromolecules* **1985**, *18*, 2305.

(5) Gunderson, J. J.; Giddings, J. C.; *Anal. Chim. Acta* **1986**, *189*, 1.

(6) Gao, Y. S.; Caldwell, K. D.; Myers, M. N.; Giddings, J. C. *Macromolecules*, **1985**, *18*, 1272.

(7) Yau, W. W; Kirkland, J. J.; Bly, D. D. *Modern Size-Exclusion Liquid Chromatography*; Wiley: New York, NY, **1979**; p225.

(8) Slagowski, E. L.; Fetters, L. J.; McIntyre, D. *Macromolecules*, **1974**, *7*, 394.

(9) Mei-Ling, Y.; Liang-He, S. *J. Liquid Chromatogr.*, **1982**, *5*, 1259.

(10) Yau, W. W. *Chemtracts-Macromolecular Chemistry*, **1990**, *1*(1), 1.

(11) Myers, M. N.; Caldwell, K. D.; Giddings, J. C. *Sep. Sci.* **1974**, *9*, 47.

(12) Giddings, J. C.; Graff, K. A.; Caldwell, K. D.; Myers, M. N. *Adv. Chem. Ser.*, **1983**, *203*, 259.

(13) Screaton, R. M.; Seemann, R. W. *J. Polymer Sci.: Part C*, **1968**, (21), 297.

(14) Williams, P. S.; Giddings, J. C. *Anal. Chem.* **1987**, *59*, 2038.

(15) Giddings, J. C.; Kumar, V.; Williams, P. S.; Myers, M. N. In *Polymer Characterization by Interdisciplinary Methods*; Craver, C. D.; Provder, T., Ed., ACS, in press.

RECEIVED April 20, 1992

SIZE-EXCLUSION CHROMATOGRAPHY: FUNDAMENTAL CONSIDERATIONS

Chapter 7

Critical Conditions in the Liquid Chromatography of Polymers

D. Hunkeler[1], T. Macko[2], and D. Berek[2]

**[1]Department of Chemical Engineering, Vanderbilt University,
Nashville, TN 37235
[2]Polymer Institute, Slovak Academy of Sciences, 84236 Bratislava,
Czechoslovakia**

The liquid chromatography of polystyrene and polymethylmethacrylate have been investigated using porous silica gel packings. It has been found that the size exclusion and interactive modes of chromatography can be combined in a single column, operating isocratically, provided the thermodynamic quality of a mixed eluent is precisely controlled. In our experiments the ratio of a nonpolar thermodynamically good-solvent (toluene) and a polar nonsolvent (methanol) were systematically varied. The calibration curves were observed to shift rapidly towards the vertical, particularly for mixtures thermodynamically much poorer than the theta-composition. Furthermore, binary mobile phase compositions were identified where the retention volume was independent of the molar mass of the polymer probe. These "limiting conditions" have been measured on polymers up to two million daltons and have been found to be largely pressure insensitive. They potentially can be applied for the molecular weight analysis of block and graft copolymers, and for polymer blends, since the separation according to size is deconvoluted from the separation by chemical composition.

The characterization of copolymers according to their molar mass and chemical composition distribution presents an important task from the point of view of both their synthesis and optimization of properties. The corresponding measurements usually include separations, most often applied by means of liquid chromatographic methods. The interactive or retentive modes of liquid chromatography (LC) or high performance liquid chromatography (HPLC) of copolymers are based on the differences in the interactions between the column packing, mobile phase and macromolecules of different composition. This leads to defined non-equalities in adsorption, absorption (partition), ionic or precipitation-redissolution processes and, consequently to the differences in the retention volumes (V_R) of particular species. The size exclusion or gel permeation chromatography (GPC) mode separates macromolecules according to their size in solution. This is governed by both the molar mass and the chemical composition of the macromolecule. The adsorption and precipitation modes, that is, high performance liquid adsorption chromatography (HPLAC) (1,2), and the high

0097–6156/93/0521–0090$06.00/0

performance liquid precipitation chromatography (HPLPC is in fact a combination of precipitation and sorption and not a classical precipitation) (3,4), are the interactive liquid chromatography procedures presently most often used for the separation of copolymers. To facilitate the control of the interactions in HPLAC and HPLPC systems, mixed mobile phases containing two or more components are used. For retention control in HPLAC eluent additives with high elution strength, that is with high desorption ability, are used. In HPLPC the precipitation-redissolution processes are governed by the addition of appropriate nonsolvents to the eluent. Temperature is another suitable parameter in HPLAC and HPLPC . Most separations in the retentive liquid chromatography of copolymers are performed with a gradient elution, where the composition of eluent is changed in either a continuous (3,4) or stepwise (5) manner according to a precise program, with the column temperature being kept constant. Unfortunately, the precision of gradient-producing liquid chromatography devices is much lower than for isocratic systems.

The conditions of HPLAC are chosen in such a way that the separation is not influenced by the molar mass of the macromolecules. That is, the mobile phase composition, temperature and column packing, nonporous, microporous or megaporous, are selected so that no size exclusion phenomena takes place within the liquid chromatography column. The fractions obtained may be further separated according to their molar mass by means of gel permeation chromatography.

The interactive liquid chromatographic separations are usually performed in the conventional elution mode in which the narrow zone of the sample is injected onto the column packing and subsequently eluted by pumping the mobile phase through the column. In the HPLAC method the stepwise elution of copolymers is also possible. Here the column packing is saturated by the adsorbed copolymer that is eluted in the following steps by the appropriate mobile phase(s) (5). The necessity of changing eluent composition in both the HPLAC and HPLPC methods presents an important disadvantage.

Although commonly used for the molar mass and molar mass distribution measurements of homopolymers, gel permeation chromatography is not directly applicable to the precise and complete characterization of copolymers, because the sizes of copolymer molecules depend on both their molar mass and their chemical composition. An interesting possibility however, involves the combination of size exclusion and interactive modes of liquid chromatography.

Multi-component Eluents

The controlled interaction of the oligomer end-groups with column packing has been shown to lead to the increased selectivity of separation as was manifested for polyethylene glycols and polypropylene glycols (6). In the case of high polymer substances, however, the influence of the end-groups is insufficient and the interactions of main chains with the column packing must be utilized. For this purpose, two- or multi-component eluents are again advantageous since they allow precise adjustments of the elution strength in connection with the non-swelling, mechanically stable, active column packings such as silica gel or porous carbon. However, in contrast with both HPLAC and HPLPC, the composition of the mixed GPC eluent is kept constant.

The addition of the second component to the eluent may substantially influence the GPC retention volumes. For example, the addition of a small amount of the appropriate substance to the eluent can either decrease or increase the retention volumes (V_R) of the highly or medium polar polymers like polymethylmethacrylate or polystyrene from the silica gel packed columns (7-9). The decrease of V_R is due to the blocking of the active sites of the silica gel by polar additive molecules such as methanol or water that suppress the **adsorption** of macromolecules. If the additive is a nonsolvent for the polymer probe it also **decreases the effective pore size**. Finally, the nonsolvent additive accumulated in the pores of the silica gel reduces the V_R due to the **partition** effects since the macromolecules prefer the solvent rich mobile phase over the nonsolvent-rich quasi stationary phase in the pores of the column packing. The opposite effect, an increase of V_R due to the addition of a second eluent, has also been observed if the additive is less polar than the eluent (e.g. n-heptane added to 2-butanone for polystyrene probes (7)). Thus, by an appropriate choice of the mixed eluent combination and composition, the calibration curves of a given polymer may be shifted in either direction while still maintaining the basic mechanism of SEC where the retention volumes decrease with increasing molar masses of the polymers.

Limiting Conditions

An interesting phenomena was recently observed by Belenkii and Gankina (10) by varying the eluent composition in a mixture of thermodynamically good solvents. The experiments involved the GPC of polystyrene on silica gel in a thin layer chromatography arrangement. They arrived at a system in which the retention volumes were independent of the molar mass of the polymer. In fact the Rf values of the polymers were as high as the Rf's of the inert low molecular weight substances. Belinkii and Gankina called these conditions "critical". Under critical conditions the macromolecules are "invisible" to the gel. In this way the gel seems to be totally permeable, independent of its pore size. Tennikov has found a similar effect in a column LC arrangement (11). Nefedov and Zhmakina (12) have added that the critical composition of an eluent, for a given polymer and sorbent, may also depend on the pressure within the system.

Belenkii and Gankina (10) proposed that at the critical eluent composition the total free energy of the process which takes place in the column is zero due to a balancing of entropic (exclusion) and energetic (adsorption) effects. This is somewhat unexpected since both the entropic and enthalpic terms change with the molecular weight of the polymer . To achieve a retention independent exclusion therefore requires the two effects to have an identical molecular weight dependence. Such a case would certainly not be expected a priori, however the elucidation of critical conditions is well documented in several polymer-binary eluent-sorbent systems, for oligomers and polymers up to 100,000 daltons (Table I). This includes the separation of mono and bifunctional poly(diethyleneglycol adipate) (19), as well as the separation of linear and cyclic macromolecules (24). The latter indicates that critical conditions can possibly separate not only according to composition and molecular weight but also by spatial structure. In this work we extend the molecular weight range to above one million daltons. We also prefer the term "limiting condition" to avoid any confusion with supercritical fluid chromatography, which is based on an entirely different premise.

Such limiting conditions are immediately attractive for the chromatography of copolymers. If the column does not see one part of the copolymer molecule the second "visible" portion can be separated as if it were in isolation. Skvortsov and Gorbunov (21) have mentioned that it is necessary for the "invisible" portion of the chain to be a free end. This may restrict the applications of limiting conditions to end blocks of block copolymers, side grafts in grafted copolymers, or polymer blends. Furthermore, the composition selective separation at limiting conditions cannot be used for the central block of a triblock copolymer, for the backbone of grafted polymers, or for random copolymers.

In our new approach for the generation of limiting conditions the eluent is defacto a nonsolvent for the polymer probe (mixture of good and poor solvents) with the polymer injected in a thermodynamically good solvent. The adsorption, and

TABLE I: Critical Conditions for Systems of Polymer-Sorbent-Eluent

Polymer	Sorbent	Eluent	Ref.
Polystyrene	Silica gel (SI)	Chloroform/tetrachloromethane, 94.5/5.5 vol%	13
Oligobutadienes	SI-100	Hexane/toluene, 85/15 vol% Hexane/dichloromethane, 76/24 vol% Heptane/MEK, 99.5/0.5 vol%	14 15
Polysulfone oligomers	SI-60	Chloroform/tetrachloromethane, 53/47 vol%	16
Polycarbonate oligomers	SI-60	Chloroform/tetrachloromethane, 30/70 vol%	16
Polybutylene-terepthalate oligomers	SI-60	THF/heptane, 80/20 vol%	16
Polypropylene glycoles	SI-100	MEK/ethylacetate, 5/95 vol%	17
Polyethylene glycoles	SI	MEK/chloroform, 35/65 vol% MEK/hexane, 92/8 vol% THF/ethylacetate, 7/93 vol% MEK/ethylacetate, 27/73 vol%	18
Polydiethylene glycoladipates	SI-100	Hexane/MEK, 8/92 vol%	19
PMMA	SI-300	Dichloromethane/acetonitrile 41.5/58.5 vol%	20
Butylmethacrylate	SI-300	Dichloromethane/acetonitrile, 90.7/9.3 vol%	21
Oligo (1,3,6-trioxocanes)	SI-C$_{18}$	Acetonitrile/water, 49.5/50.5 vol%	22
Oligocarbonates	SI-600	Chloroform/tetrachloromethane 17/83 vol%	23

THF = tetrahydrofuran, MEK = methylethylketone.

possibly also precipitation, are controlled by the **thermodynamic quality** of the eluent. This differs from the critical condition approach of Belenkii and Gankina where the **elution strength** of the eluent controls the adsorption. In our system there are still several unanswered questions. However, at this stage the following mechanism appears operative: at low levels of nonsolvent, such as water in the tetrahydrofuran-polystyrene or methanol in toluene-polystyrene or toluene-polymethylmethacrylate systems, the calibration curve shifts slightly to lower retention volumes (Figure 1) due to the influence of adsorption, partition and a reduced pore size. At higher quantities of nonsolvent, i.e. in an eluent poorer than the theta solvent (76.9% toluene for polystyrene in toluene-methanol at 25°C), the thermodynamic quality of the solvent is strongly reduced. This occurs to such an extent that as the macromolecule separates from the solvent zone (by exclusion), encounters the mobile phase and precipitates. It then redissolves as the injection zone (pure solvent) reaches the precipitated polymer. This "microgradient" process of precipitation-redissolution occurs continuously throughout the column with the polymer eluting just in front of the solvent (Figure 2). As a consequence, the column does not see the macromolecules, as in the Belenkii system, since they move with a velocity equal to the velocity of the solvent zone.

The primary differences between the Hunkeler-Macko-Berek system and the approach of Belenkii are: 1) the use of a non solvent eluent and 2) the injection of the polymer in a thermodynamically good solvent. For example, for polystyrene a limiting condition of 68/32 vol% toluene-methanol was observed at 25 °C. This is below the theta composition and therefore to maintain solubility the sample was injected in 100% toluene, a thermodynamically good solvent for the polymer probe.

Experimental

Silica gel sorbents were prepared through a polycondensation of silicic acid released from water glass by acid. These spherical, narrow pore sorbents were then modified by a chemothermal treatment in caustic solution (25) so that silica gel with an 800 Å mean pore diameter was produced. This material had a very narrow pore size distribution as measured by mercury porometry (porosimetry). Its surface area was determined by the B.E.T. method with Argon. The particles of the sorbent were sized with a zig-zag separator (Alpina) and the 10 μm fraction was selected for LC measurements. The material was not further chemically modified (silanized). It was packed into a 250 x 6 mm stainless steel column by means of a Knauer pneumatic pump operating at 30 MPa using methanol as both the slurry and the transporting liquid.

A RIDK 102 differential refractive index detector and a HPP 4001 high pressure pump (both from Laboratory Instruments Co., Prague) were employed. Pressure was measured with a custom made pressure gauge (0-25 MPa) (Institute of Chemical Process Fundamentals, CS Academy of Sciences, Prague). The refractive index and pressure signals were also recorded on a Type 185 two-pen chart recorder (Kutesz, Budapest). The capillary consisted of 100 m of 0.25 mm ID stainless steel tubing. The pressure was 1.7 MPa without the capillary and 15.0 MPa when the capillary was put inline. The injector was a PK1 model (Institute of Chemical Process Fundamentals, CSAS, Prague). Sample injections consisted of 10 μL of a polymer solution (polystyrene, polymethylmethacrylate) in a good solvent (toluene). The injected concentration was 1.0 mg/mL. The chromatographic efficiencies were tested according to conventional methods (theoretical plate number, Knox-Bristow plots) and

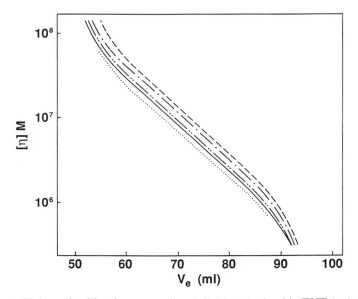

Figure 1. Universal calibration curves, log [η]M-V_e, obtained in THF (---) and the mixed eluent THF-water containing 4.5 (—·—), 7.7 (—··—), 8.2 (——), and 8.9 (·····) vol % of water.

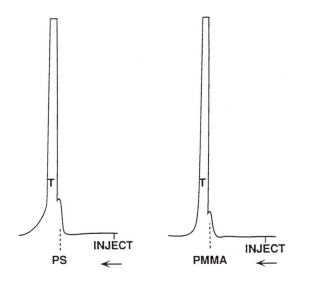

Figure 2. Chromatograms for polystyrene (PS) in 68/32 vol % toluene (T) - methanol and polymethylmethacrylate (PMMA) in 27/63 vol % toluene-methanol.

also new methods based on the pressure dependence of the preferential sorption of mixed liquids by the column packing particles (26,27).

Analytical grade solvents (toluene, methanol) were obtained from LACHEMA (Brno, CS) and used without further purification.

Polystyrene standards (polydispersity 1.06-1.20) were obtained from Pressure Chemicals Corporation (USA). Narrow molecular weight distribution polymethylmethacrylate samples were obtained from Rohm and Haas (FRG).

Results and Discussion

The calibration curves for polystyrene and polymethylmethacrylate in mixtures of toluene and methanol are shown in Figures 3 and 4. A limiting composition of 68 vol% and 27 vol% toluene are respectively observed. At such compositions of the binary eluent the retention volume is independent of molar mass over two decades in measurement. From Figure 1 (8) it can also be observed that as the thermodynamic quality of the solvent is marginally reduced the calibration curve shifts to lower retention volumes due to the combination of adsorption, partition and a reduced pore size as discussed earlier. At higher levels of nonsolvent, i.e. in eluents poorer than the theta solvent, the calibration curve is very sensitive to the composition of the mobile phase and shifts rapidly to higher retention volumes. For compositions of toluene-methanol less rich in toluene than the limiting conditions the calibration curve did not continue to shift to the right, as had been observed when the limiting conditions were caused by a balance of adsorption and exclusion (10,11). In our experiments with solvent compositions exceeding the limiting level no polymer was eluted indicating that the precipitation was irreversible over the course of the experiment, or that the redissolution by the injection zone was ineffective. The identification of adsorption domains to the right of the limiting condition (10,11) (further shifts in the calibration curve) were reported generally at lower molecular weights (10^{3-4} daltons) than employed in our experiments. The rectification of these differences requires further experimentation over a broader molecular weight range.

Figures 5 and 6 show calibration curves for polystyrene and polymethylmethacrylate in toluene-methanol at elevated pressure (15 MPa). The increase in pressure was achieved by introducing a capillary after the column. The effect of pressure did not influence the limiting composition for either polymer. However, the higher pressure experiments did correspond to a marginal change in the retention volume (10-20%) and a slightly vertical shift in the calibration curves.

The effect of pressure on the sorption equilibrium has been observed in several binary eluent-sorbent systems on both bare and modified silica gel (28). A sudden increase in pressure disturbs the equilibrium between the mobile phase and the stagnant liquid layer on the surface of the sorbent. As a result an "eigenzone" is observed in the chromatogram. A typical eigenzone for toluene-methanol-silica gel is shown in Figure 7. The height of the eigenzone is proportional to the amount of sorbent, the change in pressure and inversely related to the amount of liquid in the column and proportional to the change in pressure. In our examples, the amount of methanol in the pores is greater at higher pressures due to the preferential sorption of methanol by silica gel. This increases the retention of the polystyrene or polymethylmethacrylate which penetrate the pores due to the precipitation effect discussed herein.

Summary and Conclusions

As it has been discussed above, the separation of polymers under limiting conditions has some specific features. For example it enables one to exclude or at least to minimize the influence of different molar masses of macromolecules on the LC

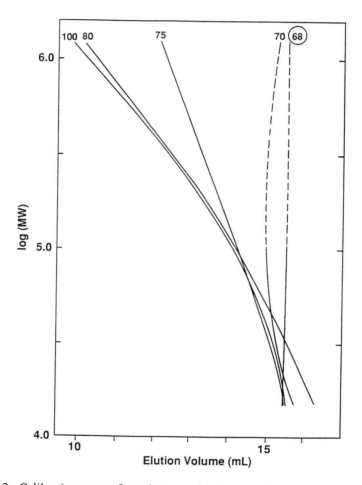

Figure 3. Calibration curves for polystyrene in mixtures of toluene-methanol at a pressure of 1.7MPa. The vol % toluene in the mobile phase is given at the top of each curve. The solid lines designate calibration curves based on polymer peaks which were identified with a refractive index detector. The dashed lines represent polymer peak retention volumes which were either not detected with the refractive index detector, or obscured by the system (toluene) peak.

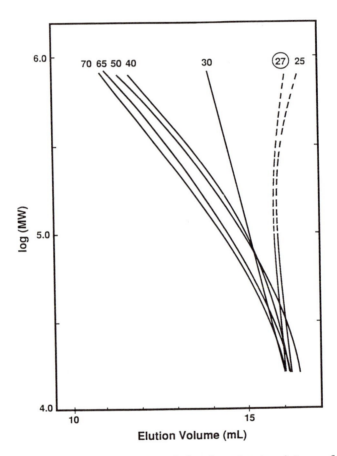

Figure 4. Calibration curves for polymethylmethacrylate in mixtures of toluene-methanol at a pressure of 1.7MPa. The vol % toluene in the mobile phase is given at the top of each curve. The solid lines designate calibration curves based on polymer peaks which were identified with a refractive index detector. The dashed lines represent polymer peak retention volumes which were either not detected with the refractive index detector, or obscured by the system (toluene) peak.

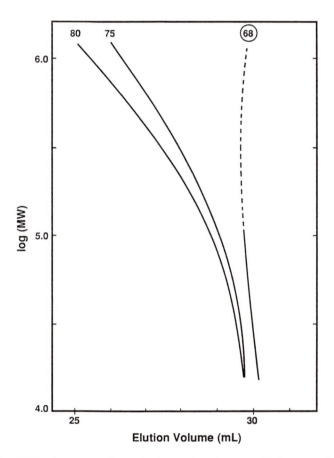

Figure 5. Calibration curves for polystyrene in mixtures of toluene-methanol at a pressure of 15.0 MPa. The vol % toluene in the mobile phase is given at the top of each curve. The solid lines designate calibration curves based on polymer peaks which were identified with a refractive index detector. The dashed lines represent polymer peak retention volumes which were either not detected with the refractive index detector, or obscured by the system (toluene) peak.

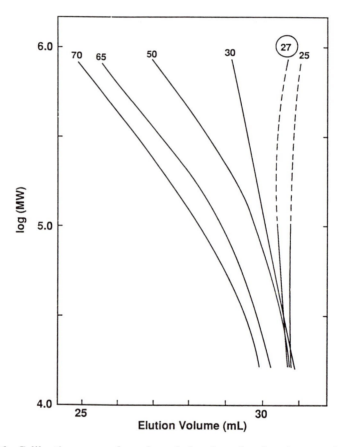

Figure 6. Calibration curves for polymethylmethacrylate in mixtures of toluene-methanol at a pressure of 15.0 MPa. The vol % toluene in the mobile phase is given at the top of each curve. The solid lines designate calibration curves based on polymer peaks which were identified with a refractive index detector. The dashed lines represent polymer peak retention volumes which were either not detected with the refractive index detector, or obscured by the system (toluene) peak.

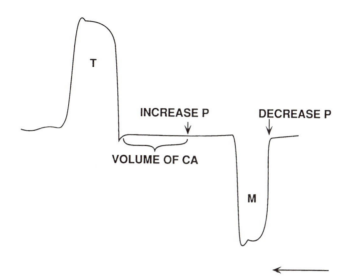

Figure 7. Eigenzones generated by decreasing the pressure (15.0 to 1.7 MPa) and increasing the pressure (1.7 to 15.0 MPa) for a toluene-methanol mobile phase with a composition of 80 vol % toluene.

separation so that other properties of both their structure and composition play a decisive role in the separation. At present a precise and reproducible method does not exist for the separation of copolymers according to molecular weight. Since the behavior of macromolecules under limiting conditions seems to be extraordinarily sensitive to the small changes in the eluent composition, temperature and pressure this approach seems very attractive for the differentiation of macromolecules. Another advantage of the separation under limiting conditions is that the separation proceeds isocratically. This excludes the problems connected with gradient elution (reequilibrating columns by long-term washing, non-suitability of some detectors, baseline instability) which render gradient procedures less sensitive than high performance liquid chromatographic methods.

The future work will focus on a comparison of the limiting conditions generated by varying the thermodynamic quality and solvent strength of the mobile phase. This will include solvent-solvent, solvent-poor solvent and solvent-nonsolvent systems with both polar and nonpolar constituents. Other additional questions which remain to be answered include the influence of the pore diameter, surface composition and preferential sorption of the silica gel, the adjustment of limiting conditions by variation in the temperature and the effect of the injected polymer concentration on the retention volume. The latter may be an important advantage of the thermodynamic quality approach since for eluents containing a combination of solvents and nonsolvents near the theta-point the retention volume is independent of the concentration of the injected polymer. However, concentration dependent retention volumes have been identified for binary eluent combinations of two thermodynamically good solvents far from the theta composition (29). Another question is the generalization of limiting conditions by correlating the shifts in calibration curves caused by eluent composition changes with macromolecular properties such as size, preferential solvation and solubility.

Acknowledgements We wish to thank Dr. I. Novák for preparing the silica gel used in this research.

Literature Cited

1. Mori, S., Uno, Y., Suzuki, M., *Anal. Chem.*, **1986**,58,303.
2. Mori, S., *J. Chromatogr.*, **1990**,507,473.
3. Glöckner, G.,*Pure and Appl. Chem.*, **1983**, 55, 1553.
4. G. Glöckner, J. Chromatogr.,384,138 (1987).
5. Konáš, M., Thesis, Institute of Macromolecular Chemistry, Czechoslovak Academy of Sciences, Prague, Czecho-slovakia, 1990.
6. Berek, D., Bakoš, D., *J. Chromatogr.*, **1974**, 91,237.
7. Berek, D., Bakoš, D., Bleha, T., Šoltés, L., *Makromol.Chem.*, **1975**, 176,391.
8. Spychaj, T., Berek, D., *Polymer,* **1979**, 20,1108.
9. Bleha, T., Spychaj, T., Vondra, R., Berek, D., *J.Polym.Sci.,Polym.Phys.Ed.*, **1983**, 21,1903.
10. Belenkii, B.G., Gankina, E.S., *J.Chromatogr.*,**1977**, 141,13.
11. Tennikov, M.B., et al., *Vysokomol. soedin.*, **1977**, B19, 677.
12. Nefedov, P.P., Zhmakina, T.P., *Vysokomol. soedin.*, **1981**, A23, 276.
13. Tennikov, M., Nefedov, P., Lazareva, M., Frenkel, S., *Vysokomol.soedin.*, **1977**, A19,657.
14. Gorshkov, A.V., Evreinov, V.V., Entelis, S.G., *Zh.fiz.khim.*, **1985**, 59,1475.
15. Gorshkov, A.V., Evreinov, V.V., Entelis, S.G., *Zh.fiz.khim.*, **1985**, 59,2847.
16. Gur'yanova, V.V., Pavlov, A.V., *J.Chromatogr.*, **1986**, 365,197.
17. Gorshkov, A.V., Evreinov, V.V., Entelis, S.G., *Zh.fiz.khim.*, **1988**, 62,490.
18. Filatova, N.N., Gorshkov, A.V., *Vysokomol. soedin.*, **1980**, A30, 953.
19. Gorshkov, A.V., Evreinov, V.V., Entelis, S.G., *Zh.fiz.khim.*, **1985**, 59,958.
20. Zimina, T.M., Kever, J.J., Melenevskaya, E.Y., Fell, A.F., J. Chromatogr., **1992**, 593,233..
21. Zimina, T.M., Kever,E.E., Melenevskaya, E.Yu., Egonnik, V.N., Belenkii, B.G., *Vysokomol.soed.*,**1991**, A33,1349.
22. Schulz, G., Much, H., Kruger, H., Wehrsted, G., *J.Liq.Chromatogr.*, **1990**, 13,1745.
23. Gorshkov,A.V., Prodskova, T.N., Guryanova, V.V., Eveinov, V.V., *Polymer Bulletin,* **1986**, 15,465.
24. Entelis, S.G., Evreinov, V.V., Gorshkov, A.V., *Adv.Polym.Sci.,* **1987**, 76,129.
25. Berek, D., Novák, I., *Chromatographia,* **1990**, 30,582.
26. Berek, D., Macko, T., *Pure and Applied Chem.*, **1989**, 61,2041.
27. Macko, T., Berek, D., *J.Chromatogr.Scr.*, **1987**, 25,17.
28. Macko, T., Chalányová, M., Berek., D.,*J.Liq.Chromatogr.*, **1986**, 9,1123.
29. Berek, D., Bakoš, D., Šoltés, L., Bleha, T., *J.Polym.Sci., Polym.Lett.Ed.*, **1974**, 12,277.

RECEIVED January 11, 1993

Chapter 8

Single-Parameter Universal Calibration Curve

R. Amin Sanayei, K. F. O'Driscoll, and Alfred Rudin

Institute for Polymer Research, University of Waterloo, Waterloo, Ontario N2L 3G1, Canada

A one parameter correlation between intrinsic viscosity, and molecular weight is presented as: $[\eta] = K_\theta \overline{M}_r^{1/2} + K' \overline{M}_w$ where $\overline{M}_w$ and $\overline{M}_r$ are weight average and radius average molecular weight of polymer respectively. K_θ and K' were estimated for PSTY, PE, PMMA, and PIB with molecular weights ranging from 10^3 to $7*10^6$ in different solvents using literature data. A parametric (logarithmic) transformation procedure was performed prior to the regression analysis. For a given polymer the value of K_θ so obtained remained constant regardless of solvent and is in excellent agreement with reported literature values. K' reflects the extent of polymer-solvent interaction. The value of K' decreased as the interaction between solvent and polymer decreased, and for a theta solvent was found to be zero.

K_θ is constant and known for many polymers; therefore, construction of a universal calibration curve using this new correlation for intrinsic viscosity leads to having a single parameter relation. For SEC system a universal calibration curve was established using polystyrene standard with molecular weight ranging from 900 to 1,800,000. The new correlation was used to convert the universal calibration to the molecular weight calibration curves for polymethyl methacrylate and polyalpha-methylstyrene using just one narrow molecular weight standard sample of each polymer. The calibration curves so obtained are in excellent agreement with the other standards of these two polymers over the entire range of molecular weight. In absence of a narrow molecular weight standard sample, this procedure may readily be extended to a single broad molecular weight sample of the second polymer.

Molecular weight distribution (MWD) is essential information for characterization of a polymer sample. This information is only attainable from size exclusion chromatography (SEC) analyses where macromolecules are separated according to their sizes in solution. In the absence of direct measurements of molecular

0097–6156/93/0521–0103$06.00/0

size of the eluant, only the distribution of macromolecules with respect to elution volume is obtained from SEC analysis. However, the MWD can be estimated by transformation of the polymer elution volumes to the corresponding molecular weights by using a universal calibration based on the premise that polymers with equal SEC elution volumes have equal hydrodynamic volumes in solution[1]. At infinite dilution it can be shown that the effective hydrodynamic volume, HV, of a dissolved random coil macromolecule is given by:

$$HV = \frac{[\eta]M}{2.5N_{Av}} \tag{1}$$

where N_{Av} is Avogadro's number and $[\eta]$ is the intrinsic viscosity of a polymer with molecular weight of M. The intrinsic viscosity in turn is related to M by the familiar Mark-Houwink-Sakurada relation[2].

$$[\eta] = K \, M^a \tag{2}$$

Since HV is a direct function of $[\eta]M$, the latter is the basis for universal calibration procedures. Consequently, if polymer species 1 and 2 have equal SEC elution volumes, then:

$$[\eta]_1 M_1 = [\eta]_2 M_2 \tag{3}$$

in the absence of concentration effects[3]. With Eq. 2:

$$\ln M_2 = \frac{1+a_1}{1+a_2}\ln M_1 + \frac{1}{1+a_2}\ln(K_1/K_2) \tag{4}$$

Thus, one can calibrate a particular SEC apparatus with polymer 1 and use the obtained molecular weight-elution volume relation for polymer 1 to measure corresponding molecular weight of polymer 2, providing that K and a are known for both polymers.

The exponent a is approximately 0.5 for low molecular weight species (e.g. polystyrene with $M < 10,000$[4]) and is independent of solvent. It is about 0.6 to 0.75 for higher molecular weight polymers. The theoretical limit is 0.8 for random coil species[5] and the value of the exponent depends on the solvent and polymer molecular weight.

Different K and a values are reported for any given polymer-solvent system[6] and experimental values of K and a are in fact inversely correlated. This is because these Mark-Houwink coefficients are obtained by fitting experimental data to Equation 2 assuming that $\log[\eta]$ is a linear function of $\log M$, as indicated by the form of the relation quoted. In fact, as shown below, the $[\eta] - M$ relation is not exactly linear over a wide enough molecular weight range. Thus, experimental data fits are actually chords to a shallow curve and the particular values obtained will depend on the experimental range of M. In addition, use of linear least squares data correlation gives heavy weights to the particular extreme molecular weight values, and it will vary with the samples used by different investigators. For these reasons, tabulations of K and a values normally list the molecular weight ranges of the samples used[6, 7].

Universal calibration uses two sets of K and a values (Eq. 4) and rather severe errors can be introduced into the estimation of the molecular weight of the sample of interest(M_2) if the Mark-Houwink coefficients are not constant over the molecular weight range.

In order to demonstrate the changes in K and a values with M we gathered data for the measured intrinsic viscosity of polystyrene (PSTY) in toluene and polymethyl methacrylate (PMMA) in benzene over wide range of molecular weights (10^3 to $7 * 10^6$) from published work. Only monodisperse or well fractionated samples were considered. Plots of $\log[\eta]$ versus $\log M$ for PSTY in toluene and PMMA in benzene, shown in Fig. 1 and 2 respectively, display clearly that K and a are not constant with molecular weight.

One really needs different sets of K and a values for different molecular weight ranges in order to make accurate estimates of MWDs. This would be an awkward expedient; however, we present a convenient alternative method. In this paper we discuss a new correlation between $[\eta]$ and MW which has only a single adjustable parameter and show how this new correlation can be used to convert the UCC to a calibration curve based on a single standard sample.

A New Correlation Between $[\eta]$ and MW

Through simple theoretical modeling we postulate the following correlation for molecular weight dependence of intrinsic viscosity.

$$[\eta] \;=\; K_1 M^{1/2} + K_2 M \tag{5}$$

Details of the model and its derivation will be published separately[8]. K_1 and K_2 were evaluated for PSTY, PMMA, polyethylene (PE),and polyisobutene (PIB) with molecular weights ranging from 10^3 to $7 * 10^6$ in different solvents. A parametric transformation of data prior to regression was found to be necessary due to the fact that the error structure of data was heteroscedastic (i.e. the error variance increases with an increase in M) [8]. We found that the logarithmic transformation improved the error structure to a satisfactory level. The logarithmic transformation of Eq. 5 is

$$\log([\eta]) \;=\; \log(K_1 M^{1/2} + K_2 M) \tag{6}$$

The estimates of K_1 remained constant regardless of solvent, and numerically were equal to K_θ for a given polymer. The estimate of K_2 decreased as the interaction between solvent and polymer decreased, with K_2 approaching zero for a theta

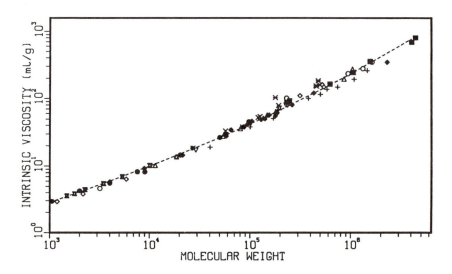

Figure 1 Intrinsic viscosity vs. molecular weight for PSTY in toluene; symbols are from 11 different literature cites, dotted line is Eq. 5

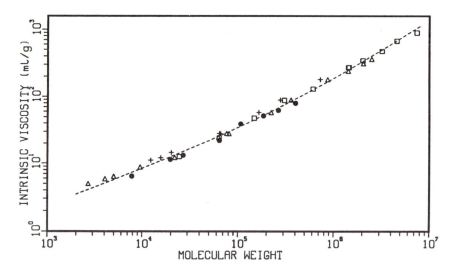

Figure 2 Intrinsic viscosity vs. molecular weight for PMMA in benzene; symbols are from 5 different literature cites, dotted line is Eq. 5

solvent. The estimates of K_1 and K_2 are listed in Table 1, along with literature values of K_θ for the above polymers. In order to eliminate redundancy of symbols Eq.5 can be presented in an alternative form of:

$$[\eta] = K_\theta M^{1/2} + K'M \tag{7}$$

When the polymer sample is polydisperse Eq. 7 leads to Eq. 8:

$$[\eta] = K_\theta \overline{M}_r^{1/2} + K'\overline{M}_w \tag{8}$$

where

$$\overline{M}_r \equiv \left[\sum w_i M_i^{1/2}\right]^2 \quad and \quad \overline{M}_w \equiv \sum w_i M_i$$

The weight average molecular weight, $\overline{M}_w$, and the radius (of gyration) average molecular weight, $\overline{M}_r$, are intrinsic properties of a given polymer sample regardless of solvent and temperature. The physical meaning of $\overline{M}_w$ is well understood. The physical interpretation of $\overline{M}_r$ is that $\overline{M}_r$ is a molecular weight average which corresponds to the average radius (of gyration) of the polymer coil, $< r_p >$, under θ conditions.

$$\overline{M}_r = M_0 \sigma < r_p >^2 = M_0 \sigma \left[\sum w_i < r_i >\right]^2 \tag{9}$$

where M_0 is molecular weight of repeating units and σ is a proportionality constant and depends on the characteristic ratio of a polymer chain.

The radius average molecular weight $\overline{M}_r$ for a given sample can be measured directly from the intrinsic viscosity of the sample in a θ solvent, provided that K_θ is known.

$$\overline{M}_r = \left(\frac{[\eta]_\theta}{K_\theta}\right)^2 \tag{10}$$

In Figs. 1 and 2 the dotted lines are the prediction of Eq. 5 using the corresponding K_1 and K_2 values from Table 1.

Universal Calibration Curve

In SEC analysis polymers appear in the eluent in reverse order of their hydrodynamic volumes in the particular solvent. The hydrodynamic volume or size of solvated polymer is proportional to $[\eta]M$. When the intrinsic viscosity and molecular weight of polymer standards are known the UCC can readily be constructed.

$$HV_i = M_i \times [\eta_i] = f(V_i) \tag{11}$$

where V_i is the elution volume of corresponding hydrodynamic volume (HV_i). In many cases $[\eta_i]$ is not known and it is estimated from a $[\eta] - M$ relation for polymer standard of known M. The MHS relation (Eq. 2) is very often used as mentioned; thus, the hydrodynamic volume of a polymer would be:

$$HV_i = KM_i^{a+1} \tag{12}$$

Table 1: Estimates of K_1 and K_2 in (ml/g), † K_θ value from literature [6]

polymer/solvent	$K_1 * 10^2$ $(K_\theta * 10^2)$	$K_2 * 10^5$ $(K' * 10^5)$	K_θ† $* 10^2$
PS/THF	8.512 ± 0.50	17.42 ± 1.36	8.2 ± 0.5
PS/Toluene	8.518 ± 0.41	14.84 ± 1.10	
PS/MEK	8.500 ± 0.34	3.68 ± 0.52	
PS/Cyclohexane	8.395 ± 0.12	0.087 ± 0.14	
PS/TCB $145°C$	7.420 ± 0.61	15.04 ± 0.15	
PE/Tetralin	24.90 ± 2.8	$160.0 \pm 17.$	28 ± 5
PE/1-Cl-naphthalene	29.10 ± 5.4	10.99 ± 7.7	
PE/diphenyl ether	29.31 ± 2.7	1.64 ± 4.75	
PE/TCB	32.88 ± 0.5	39.46 ± 0.25	
PMMA/Benzene	7.302 ± 0.44	11.33 ± 1.14	7.2 ± 2
PIB/Cyclohexane	12.17 ± 1.03	31.88 ± 3.85	10.7 ± 0.5

In practice values of K and a which give an accurate $[\eta]$ along the mid range of MW would be used. As MW falls below or exceeds this range the predictions of Equation 12 become erroneous.

On the other hand, the hydrodynamic volume based on the alternative relationship (Eq. 7) is:

$$HV_i = K_\theta M_i^{3/2} + K' M_i^2 \tag{13}$$

According to this model the polymer hydrodynamic volumes consists of two components: molecular weight to the three halves power, $M^{3/2}$, and molecular weight to the second power, M^2. K_θ is constant and dependent only on the structure of the polymer and, to some extent on temperature. The value of K' decreases as the solvent-polymer interaction decreases and becomes zero for a θ solvent.

Experimental

The SEC measurements were performed with a chromatographic apparatus equipped with a UV detector using 254 nm cut off filter, and a differential refractometer (DRI). The detectors were linked to a microcomputer through an interface for data acquisition. A set of PL-gel columns (Polymer Laboratories, Shropshire, UK) with pore size of 10^5, 10^3, 500, and 100Åand bead size of 10μ were used at the operating temperature of $25°C$. The HPLC grade tetrahydrofuran (THF) was utilized as eluent and the flow rate was kept at 1.0 ml/min.

The polystyrene standards were obtained from Pressure Chemical Co. (Pittsburgh, PA. USA). The PMMA standards were received from Polymer Laboratories Ltd. (Shropshire, UK). The polyalpha-methyl styrene (PASTY) standards were from Polysciences Inc. (Warrington, PA. USA)

Results and Discussion

PSTY standards were used to construct the UCC since the values of K_θ and $K^{'}$ for PSTY in THF were determined from the literature (Table 1). In doing so we found that the UCC can be represented best by a cubic polynomial for 15 standard samples with M of 900 to 1,800,000. Fig. 3 displays hydrodynamic volume of PSTY versus elution volume.

The elution volume of a single PMMA standard with M of 185000 was measured using the same system. Subsequently the intrinsic viscosity of this sample was determined from its hydrodynamic volume using the UCC. The value of K_θ for PMMA is known; thus, $K^{'}$ was readily obtained from Eq. 7.

$$K^{'} = \frac{[\eta] - K_\theta M^{1/2}}{M} \tag{14}$$

Knowing K_θ and $K^{'}$ values ($K_\theta = 7.3 * 10^{-2}$ and $K^{'} = 1.12 * 10^{-4}$) we were able to convert the UCC to a M calibration curve for PMMA. This calibration curve is the dashed line in Fig. 4. In order to ensure that the calibration curve is accurate over the entire range, the elution volume of 10 other PMMA standards were measured. The results are also plotted in Fig. 4. An excellent agreement exists between the calibration curve from the single point measurement and the PMMA standards over a wide range of molecular weight (from $3,800$ to $1,600,000$).

The same procedure for PASTY was adopted. In the literature K_θ is reported to be $7.4 * 10^{-2}$ and the estimate of $K^{'}$ is $13.5 * 10^{-5}$ for PASTY standard with M of $139,000$. Fig 5 shows the calibration curve from single point measurement and the experimental result from the analysis of PASTY standards with M ranging from $19,500$ to $700,000$.

If a standard sample with narrow MWD is not available the molecular weight calibration can be accomplished by using a single broad MWD sample($\overline{M}_w/\overline{M}_n > 2.5$). $K^{'}$ can be estimated from the intrinsic viscosity in eluent and the SEC chromatograph of this sample providing K_θ is being known. Molecular weight of the new polymer (M^*) is related to the hydrodynamic volume from Eq. 13 as the following:

$$HV_i = K_\theta M_i^{* 3/2} + K^{'} M_i^{* 2} \tag{15}$$

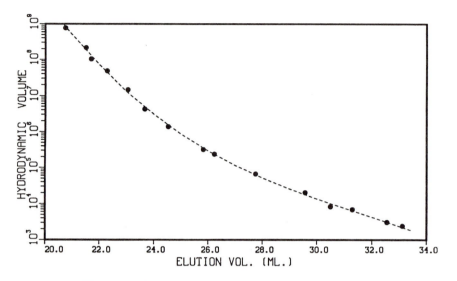

Figure 3 Hydrodynamic volume vs. elution volume

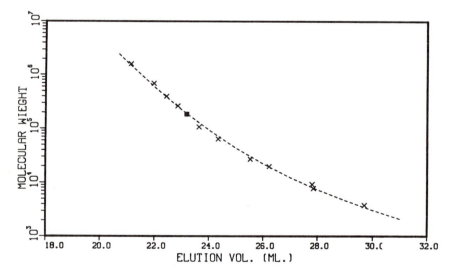

Figure 4 (– – –) PMMA calibration curve from single point (filled square), × PMMA standards

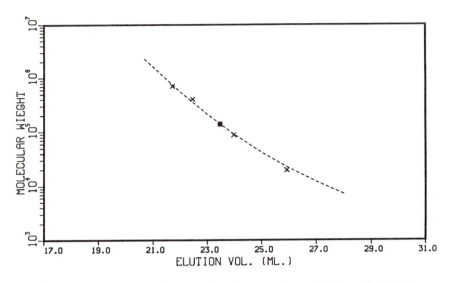

Figure 5 (– – –) PASTY calibration curve from single point (filled square), × PASTY standards

On the other hand, the intrinsic viscosity of the polymer sample is the weight average value of the intrinsic viscosities of its components in a given solvent. Thus:

$$[\eta] \ = \ \sum w_i [\eta_i] \ = \ \sum w_i (K_\theta M_i^{*\,1/2} + K' M_i^*) \tag{16}$$

where w_i is the weight fraction of macromolecules with molecular weight of M_i^*. A simple computer calculation can be done using known K_θ and experimental values of $(w_i, \ HV_i)$ together with assumed values of K'. From such a calculation one can (with Eq.15) determine M_i^* and (with Eq. 16) the optimal value of K' which gives the correct value of $[\eta]$. Note that HV_i is known from universal calibration and $[\eta]$ is experimentally measured.

Conclusion

The new correlation (Eq. 7) was used to model the intrinsic viscosities of PSTY, PMMA, PE, and PIB in different solvents over molecular weights ranging from 1,000 to 7,000,000. K_θ and K' were estimated after a parametric transformation procedure had been performed on the model. Values of K_θ so obtained are in excellent agreement with the reported literature value for these polymers, indicating that the molecular weight dependence of the intrinsic viscosity is well represented by a single adjustable parameter model.

The new model can be employed to estimate the hydrodynamic volume of solvated polymer. K_θ is constant and known for many polymers and K' indicates the extent of polymer-solvent interaction. Therefore, use of this model leads to a single parameter universal calibration curve. Molecular weight calibration curves for PMMA and PASTY were obtained from universal calibration curve using only one standard sample of each polymer. The molecular weights obtained from these calibrations are in excellent agreement with the standard samples.

Acknowledgement

Support of this research by the Natural Sciences and Engineering Research Council of Canada and by the Ontario Center for Materials Research is gratefully acknowledged.

References

1. Benoit H., Rempp P., and Grubisic Z., *J. Polym. Sci.*, Phys. Ed. 5, 753, (1967)
2. Rudin A., *The Elements of Polymer Science and Engineering* Academic Press, New York City, (1982)
3. Mahabadi H.KH., and Rudin A., *Polymer J.* , 11, 123, (1979)
4. Yamakawa H., *Macromolecules* 22, 3419, (1989)
5. Flory P.J. and Fox T.G, *J. Amer. Chem. Soc.* , 73, 1909 (1951)
6. Brandrup, J., and Immergut, E.H., *"Polymer Handbook "* 3nd Ed., John Wiley and sons, Inc. New York City (1989)
7. *American Society for Testing and Materials (ASTM)* 8-3, D-3593 (1991)
8. Amin Sanayei R. and O'Driscoll K.F., ms in preparation
9. Kurata M. and Stockmayer W. H., *Adv. High Polym. Sci.* 3, 198, (1961)

RECEIVED April 20, 1992

Chapter 9

Specific Refractive Index Increments Determined by Quantitative Size-Exclusion Chromatography

Rong-shi Cheng and Shi-lin Zhao

Department of Chemistry, Nanjing University, Nanjing 210008, People's Republic of China

A simple method of measuring specific refractive index increment is presented. The method utilizes conventional SEC apparatus equipped with a differential refractive index detector and normal operating procedure. The measuring principle is based upon fundamental rules of absolute quantitation of SEC detector responses. The applicability of the method is examined by measurements with various polymer-solvent systems. The partial specific volume and bulk refractive index of polymers can be estimated also by the proposed method.

Knowledge of the specific refactive index increment (s.r.i.i.) is of the utmost importance for the evaluation of static light scattering measurements and for controlling the sensitivity of differential refractive index detector of size exclusion chromatographic or liquid chromatographic equipments. The s.r.i.i. is usually determined with a specially designed differential refractometer or interferometer (1,2). The differential refractometric detector of normal SEC or HPLC apparatus has also been used separately as a unit equipment for measuring s.r.i.i. under static operating conditions (3,4,5). Recently in this laboratory the fundamental rules of absolute quantitation of SEC have been clarified and by which the absolute chain length distribution and molecular weight dependence of s.r.i.i. of oligomeric polystyrene were measured with conventional SEC apparatus and normal chromatographic operating procedure (6,7) The principle for measuring s.r.i.i. by quantitative SEC is extended to whole polymer and any solute in the present paper and examined with various polymer-solvent systems.

Measuring Principle

The response height, H, monitored by a differential refractive index detector of SEC is proportional to the refractive index difference, Δn, between the solution and the solvent in the detector cell. For a mono-disperse solute dissolved in a single solvent, the refractive index difference equals the product of s.r.i.i. and concentration of the solution

$$\Delta n = (dn/dC)\ C \qquad\qquad (1)$$

Since both the response heights and concentrations are functions of elution volume, V, we have

$$H(v) = k\ (dn/dC)\ C(v) \qquad\qquad (2)$$

Where k is a instrumental constant related to the chosen optical and electronic gain factor of the differential refractive index detector. Owing to the instrumental spreading effect of SEC, the maximum concentration of the solution in the detector cell is usually unknown and always less than the concentration of the injected solution, C_{inj}. The total area of the experimental chromatogram obtained by integrating Eq.2 with the assumption that dn/dC is constant

$$A = \int H(V)\ dV = k\ (dn/dC)\ \int C(V)\ dV = k\ (dn/dC)\ w \qquad (3)$$

is proportional to the mass of injected solute, w, which in turn is a known quantity and equals to the product to C_{inj} and the volume of the injected solution V_{inj}

$$w = \int C(V)\ dV = C_{inj}\ V_{inj} \qquad\qquad (4)$$

Designating the product of the instrumental constant and s.r.i.i. as the response constant of the DRI detector for the injected solute, K, we have

$$K = k\ (dn/dC) = A\ /\ w \qquad\qquad (5)$$

which is the ratio A/w and readily determinable by injecting known amounts solute into the SEC column and measuring the total area of the chromatograms obtained. The variation of the injected solute mass can be achieved more conveniently by changing the injected solution volume with known concentration, but injecting a given volume of sample solution with varying concentration should give identical results.

If a solute with known s.r.i.i. is chosen as standard sample, and its response constant

$$Ks = (A/w)s = k \ (dn/dC)s \qquad (6)$$

is also determined under the same experimental conditions (gain factor, flow rate, recorder chart speed etc.), since the instrumental constant k of given DRI detector is unique regardless what solvent is used, combining Eq. 5 and 6 we have

$$(dn/dC) = (K/Ks) \ (dn/dC)s \qquad (7)$$

by which the s.r.i.i. of the testing sample is readily evaluable from the detector response constant and the s.r.i.i. value of the standard.

Selection of Standard

Aqueous salt or sugar solutions with known refractive index usually have been chosen a standards in traditional differential refractometry (1), but in nonaqueous liquid chromatographic system they are not suitable for present purpose. Tetrahydrofuran solution of benzene was first selected as primary standard and its s.r.i.i. was calculated according to Lorentz-Lorenz formula (1)

$$(dn/dC) = \bar{v}[(n^2-1)/(n^2+2)-(n_0^2-1)/(n_0^2+2)][(n_0^2+2)^2/(6n_0)] \qquad (8)$$

in which n and n_0 re the refractive index of solute and solvent respectively, $\bar{v}$ is the partial specific volume of the solute and approximates to the reciprocal of its density ρ. The literature values of n = 1.5020, n_0 = 1.4066 and ρ = 0.8737 were used for the calculation. The calculated s.r.i.i. value of benzene in tetrahydrofuran (0.1055) was regarded as (dn/dC)s of primary standard. Since both benzene and tetrahydrofuran are volatile liquids with low boiling point, ascertainment of the exact concentration of the standard solution even by weighing is not a easy task. For overcoming this shortness selecting a nonvolatile solute with known s.r.i.i. as standard may be better. Therefore, toluene solution of polystyrene was taken as reference standard, using its literature s.r.i.i. value 0.111 as (dn/dC)s to evaluate the s.r.i.i. of testing samples.

Specific Refractive Index Increment

For examining the applicability of the method proposed in the preceding paragraphs, the DRI detector response constants of a number of polymer-solvent systems were measured by a conventional SEC apparatus operating with normal procedure. The systems studied included polystyrene and polydimethylsiloxane in tetrahydrofuran, toluene and chloroform, octamethylcyclotetrasiloxane (D4) in toluene and benzene in tetrahydrofuran, in which benzene in tetrahydrofuran and polystyrene in toluene were taken as the primary and reference standard respectively.

The relationship between the total area of experimental chromatogram and
the injected solute weight of these systems were shown in Fig. 1. All of
which could be represented by a straight line passing through origin.
The average value of the quotient A/w was taken as the experimental response constant for each system. The s.r.i.i. were calculated according
to Eq.6 from the experimental response constants of the sample and the
standard. The results obtained are listed in Table 1. The data in the
Table indicate that the s.r.i.i. calculated with primary and reference
standard are both in accord with the literature values. This results
shows that the conventional SEC apparatus widespreadly distributed in
analytical and polymer laboratories could be used directly to measure
s.r.i.i. of any solute conveniently with normal operating procedure by
the principle proposed above.

Two additional comments should be noted.

1. Usually white light is used for conventional DRI detector of SEC
apparatus instead of monochromatic light source required for s.r.i.i.
determination. Under present circumstances with the assumption that the
dependence of s.r.i.i. of all solutes on wave length are nearly the same, we may consider that the calculated s.r.i.i. of the testing samples
are the values under the specified wave length same as that of the standard.

2. In the normal SEC process any low molecular weight impurities in the
sample solution are separable and detectable. This is an additional advantage over the traditional static differential refratometry.

Partial Specific Volume and Refractive Index

The s.r.i.i. may be calculated by the theoretical formula of Lorentz-
Lorenz as Eq.8 or by the more simple empirical formula of Gladstone-Dale
as

$$(dn/dC) = \bar{v} \, (\, n - n_0 \,) \qquad\qquad (9)$$

These two formulas suggest that the partial specific volume of the solute could be estimated from s.r.i.i. measurements (8,9).

Assuming the partial specific volume of a given solute in different solvents are nearly the same, then both the partial specific volume and
bulk refractive index of the solute could be estimated simultaneously by
measuring s.r.i.i. in several solvents with different refractive index.
The quantitive SEC provides a convenient means to do this. Combining
Eq. 5 with Lorentz-Lorenz (LL) formula of Eq. 8 and Gladstone-Dale formula (GD) of Eq. 9, we have

$$\text{LL:} \quad 6n_0 \, K/(n_0^2 + 2)^2 = k\bar{v}(n^2 -1)/(n^2 +2) - k\bar{v}(n_0^2 -1)/(n_0^2 +2) \qquad (10)$$

$$\text{GD:} \qquad K = k\bar{v}n - k\bar{v}n_0 \qquad\qquad (11)$$

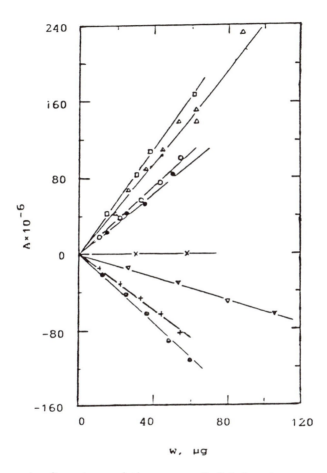

Figure 1. Dependence of the area of D.R.I. chromatograms
for various solute-solvent systems on the injected
sample amounts.

☐ PS-THF
△ PS-Chloroform
○ PS-Toluene
● Benzene-THF
▽ PDMS-Chloroform
+ PDMS-Toluene
× PDMS-THF
◑ D4-Toluene

Table 1 D.R.I. Response Constants and Specific Refractive
Index Increments of Various Solute-Solvent Systems

Solute	Solvent	$K \times 10^{-6}$	dn/dc, SEC		dn/dc Literature Data
			Φ H-THF as Standard	PS-Toluene as Standard	
Benzene	THF	1.59	–	0.102	0.1055 *
PS	Toluene	1.73	0.114	–	0.111 * * [8]
	THF	2.80	0.186	0.180	0.189 [1 0]
	Chloroform	2.43	0.161	0.156	0.169 [1 1]
PDMS	Toluene	-1.47	-0.098	-0.094	-0.093 [1 2]
	THF	0	0	0	–
	Chloroform	-0.60	-0.040	-0.038	–
D4	Toluene	-1.76	-0.116	-0.113	–

*   Primary standard, calculated value.
* * Reference standard, literature data.

Plots of $6n_0 K(n_0^2 + 2)^2$ versus $(n_0^2 - 1)/(n_0^2 + 2)$ and K versus n_0 should be linear and the partial specific volume and refractive index of the solute could be estimated from the intercept I and slop S of the straight line drawn for these two plot as

$$\bar{v} = S_{LL}/k = S_{GD}/k \tag{12}$$

$$n = [(1+2I_{LL}/S_{LL})/(1-I_{LL}/S_{LL})]^{\frac{1}{3}} = I_{GD}/S_{GD} \tag{13}$$

in which the instrumental constant k may evaluated from the response constant of the standard with Eq. 6.

Such plots for polystyrene and polydimethylsiloxane in tetrahydrofuran, chloroform and toluene are shown in Fig. 2. The refractive indexes of the solvents are taken from literature (1). which are 1.4066, 1.4446 and 1.4980 respectively. The instrumental constant k calculated from the response constant and s.r.i.i. of the primary standard equals to 15.1×10^6 The intercepts and slopes of the straight lines in Fig. 2 were obtained by linear regression and from which the partial specific volume and refractive index of the polymer are estimated by Eq. 12 and 13. The results obtained are listed in table 2. The estimated values are close to that reported in the literature.   Therefore, quantitative SEC also provides a convenient means to estimate these two important physical constants.

Experimental

A Waters model 244 liquid chromatography apparatus consisted of a differential refratometer R401, a ultrastyragel column with porosity 500 A. a universal injector U6K and a data module waters DM 730 was used. All measurements were carried out at a flow rate of 0.5 ml/min., a tem- perature of 25 ℃. and detector sensitivity of ×8. Tetrahydrofuran, to- luene and chloroform were successively used as eluents. The area of the chromatograms were recorded and printed by the data module. The concen- tration of the injected sample solution was ascertained by weighing the solute before addition of the solvent and weighing again the solution after complete dissolution. The weight concentration in g./g. unit was converted to g./ml. unit with the known density-concentration relations of each sample solution. The concentrations of the injected solution were in the range of 1 to 3 mg./ml. and volume of the injected solution were in the range of 10 to 100 μ 1.

Polystyrene with weight average molecular weight of 9.26×10^6 was supp- lied by Waters. The polydimethylsiloxane and octamethylcyclotetrasilo- xane were prepared in this laboratory.

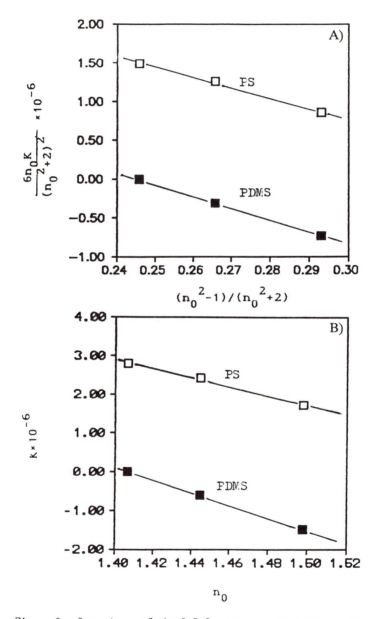

Figure 2.   Dependence of the D.R.I. response constants on the
            refractive indexes of solvents.
            A: Plots according to Lorentz-Lorenz Eq. 10.
            B: Plots according to Gladstone-Dale Eq. 11.

Table 2. The Partial Specific Volume and Refractive
Index Estimated from S.R.I.I. Measurements

Polymer	$\bar{v}$			n		
	LL	GD	Literature Data	LL	GD	Literature Data
PS	0.892	0.783	0.89-0.94	1.635	1.646	1.59
PDMS	1.033	1.068	1.02-1.05	1.407	1.407	1.43

Acknowledgments: The project was supported by the National Natural
Science Fundation of China.

Literature Cited
1. Huglin, M. B., in"Light Scattering From Polymer Solutions", Huglin,
   M. B., Ed., Academic Press, N.Y. 1972, Chap. 6
2. Huglin, M. B., in"Polymer Handbook", Brandrup, J. and Immergut E. H.
   Ed., John Wiley & Sons, N.Y. 1975, 2nd Edition, IV-267.
3. Candau, F., Francois, J. and Benoit, H., Polymer, 15, 626 (1974).
4. Soria, V., Llopis, A., Cerda, B., Campos, A. and Fiquerueto, J. E.,
   Polymer Bulletin, 13, 83 (1985).
5. Philipps, T. H. and Borchard, W., Eur. Polym. J., 26, 1289 (1990).
6. Yan, X.-H., Yu, X.-H. and Cheng, R.-S., Acta Polym. Sinica, 1989(5),
   558.
7. Cheng, R.-S. and Yan, X.-H., Acta polym. Sinica, 1989 (6), 647.
8. Bodmann, O., Makromol. Chem., 122, 196 (1969).
9. Heller, W., J. Polym. Sci., A2, 4, 209 (1966).
10. Schulz. G. V. and Baumann, H., Makromol. Chem., 114, 122 (1968).
11. Kotera. A., Iso, N., Senuma, A. and Hamada, T., J. Polym. Sci., A2,
    5, 277 (1967).
12. Nilson, R. and Sundelof, L. O., Makromol. Chem., 66, 11 (1963).

RECEIVED April 20, 1992

Chapter 10

Copolymer Characterization Using Conventional Size-Exclusion Chromatography and Molar-Mass-Sensitive Detectors

F. Gores and P. Kilz

PSS Polymer Standards Service GmbH, P.O. Box 3368, D–6500 Mainz, Federal Republic of Germany

Several types of block copolymers, comb-shaped polymers and random copolymers were investigated using multiple detection SEC, on-line multi-angle laser light scattering and on-line viscometry. Multi-detector SEC can measure chemical composition distribution by detector calibration, which is very useful to understand copolymerization processes. The validity of comonomer composition agrees very well with NMR and GC data for different copolymer types and comonomers.

Copolymer molar masses are calculated using homopolymer calibration curves and compositional information. In the case of block copolymers a good agreement is generally found for all methods employed. The calculated copolymer calibration curve matches that measured by light scattering for a star-shaped block copolymer.

In the last 30 years size-exclusion chromatography (SEC) has become a major tool for polymer characterization. Molar mass averages and molar mass distributions can be measured based on a calibration for the polymer under investigation. The analysis of copolymers is more difficult, since SEC retention is governed by molecular size and not molar mass.

This paper describes a reliable and rapid method for the analysis of copolymers by size-exclusion chromatography (SEC). The properties of copolymers mainly depend on the choice of comonomers, molecular weight and composition. In the case of homopolymers the molecular weight distribution determines many important properties. Additionally, the knowledge of composition distribution for copolymers is most important, since it influences physical properties, e.g. chain dimensions and rheological data, to a great extend.

Three on-line methods are used to try to characterize copolymers by SEC with respect to molar mass and composition distribution:
1. Conventional SEC utilizing multiple detection

2. On-line analysis of SEC fractions with a multi-angle laser light scattering (MALLS) detector
3. On-line viscometry

An on-line combination of these methods allows the independent measurement for each SEC slice of chemical composition (method 1), M_w (method 2) and $[\eta]$ (method 3). An estimate of the number average molecular weight for the whole copolymer is also possible by employing Goldwasser's *(1)* approach for the integration of the intrinsic viscosity distribution.

Theory

SEC separation of copolymers is generally more complex than SEC characterization of homopolymers. This is due to the fact that a copolymer shows a molar mass distribution as well as a comonomer distribution. Since a polymer is separated according to molecular *size*, SEC may yield fractions that can be polydisperse in molar mass *and* chemical composition (cf. Figure 1). For example, in the case of styrene/MMA copolymers of the same molar mass, retention increases in the following order: poly(styrene-<u>r</u>-MMA), homo-polystyrene, poly(styrene-<u>b</u>-MMA) and homo-PMMA *(2)*.

SEC with multiple concentration detectors. Since SEC separation is based on hydrodynamic volume rather than the molecular weight of the polymer, calibration data are only valid for polymers of identical structure. This means that polymer topology (e.g. linear, star-shaped, comb, ring or branched polymers), copolymer composition and chain conformation (isomerization, tacticity, etc.) determines the apparent molecular weight.

The main problem of copolymer analysis is the calibration of the SEC instrument for copolymers with varying comonomer compositions. But even if gross composition is constant for the sample under investigation, second order chemical inhomogeneity has to be taken into account, i.e. composition generally varies with molecular weight.

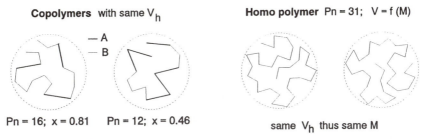

Figure 1. Comparison of hydrodynamic volume for homo and copolymers and its dependence on chain length and composition.

Several attempts have been made to solve the calibration dilemma. Some are based on the universal calibration concept which has been extended for copolymers (3), (4) another approach to copolymer calibration is multiple detection (5).

The advantage of multiple detection can be seen in its flexibility and yielding the composition distribution as well as molecular weights for the copolymer under investigation.

Determination of chemical composition. Chemical heterogeneity of copolymers can be measured by detector calibration. Output of any concentration detector d, U_d, is proportional to the concentration of comonomers k, incorporated in the copolymer. The detector response factor, f_{dk}, is determined by injection of e.g. homopolymers or model compounds from the detector response and the injected mass.

$$U_d = \sum_d f_{dk} \cdot c_k \tag{1}$$

where f_{dk} is the response factor for comonomer k at detector d

The determination of k components in a copolymer requires a SEC instrument having k independent detectors in order to solve the (k x k) matrix and to calculate the absolute concentrations $c_{dk}(V)$ of all comonomers in each detector cell.

In the case of a binary copolymer the weight fraction, w_a, of comonomer A is then given by:

$$W_a(V) = \left[1 + \frac{[U_1(V) - \dfrac{f_{1b}}{f_{2b}} \cdot U_2(V)][f_{1a} - \dfrac{f_{1b}}{f_{2b}} \cdot f_{2a}]}{[U_1(V) - \dfrac{f_{1a}}{f_{2a}} \cdot U_2(V)][f_{1b} - \dfrac{f_{1a}}{f_{2a}} \cdot f_{2b}]} \right]^{-1} \tag{2}$$

The copolymer detector trace can be separated using the individual comonomer concentrations according to:

$$h_d(V) = \sum_k c_k \cdot h_{dk}(V) \tag{3}$$

Molar Mass Moments of Copolymers. These were calculated as follows:

$$\bar{M}_{nc} = \frac{\sum_i \sum_k h_{ik} \cdot M_{ik}}{\sum_i \sum_k h_{ik}} \qquad \bar{M}_{wc} = \frac{\sum_i \sum_k h_{ik} \cdot M_{ik}^2}{\sum_i \sum_k h_{ik} \cdot M_{ik}} \qquad D_c = \frac{\bar{M}_{wc}}{\bar{M}_{nc}} \tag{4}$$

This double summation is mathematically similar to Runyon's method (5) to calculate copolymer molar masses, M_c, from homopolymer calibration curves and compositional information w_k (cf. Figure 2):

$$lg\, M_c = \sum_k w_k \cdot lg\, M_k \tag{5}$$

A pure combination of homopolymer calibration curves, as given in eq. (5), may lead to physically not valid results, e.g. copolymer M_w values smaller than M_n values or even negative values for M_w or M_n. This abnormal behavior is primarily caused by detector noise especially at the onset of the polymer peak. Model calculations also revealed that drastic changes in copolymer composition will cause such behavior. This would lead to positive slopes in a copolymer calibration curve calculated using eq. (5). Copolymer molar mass moment calculations using the double summation given in eq. (4) overcome such influences.

Copolymer analysis using the PSS COPO SEC software package is based on this modified multi-detection method first reported by Runyon et al. *(5)*. Molecular weight and composition information is obtained in the same SEC run without any special sample preparation necessary.

As was already shown by Benoit et al. *(6)*, molar masses of the copolymers are accurate, if segment-segment interactions are negligible. The precision of the compositional information is not affected by polymer topology, however. Deviations in the comonomer ratio are only conceivable, if the detected property is dependent on the environment. This is the case if neighbor-group effects exist. However, the possibility of electronic interactions causing such deviations is very low, since there are too many chemical bonds involved. Other types of interactions especially those which proceed across space (e.g. charge-transfer) may influence composition accuracy.

It has been shown that no specific calibration is needed if the copolymers have block structures *(7)*. Different copolymer topologies like comb-shaped or star-shaped polymers may require special calibration or further characterization (e.g. by on-line MALLS or viscometry) in order to give accurate molecular weights.

Utilization of On-line Multi-Angle Laser Light Scattering. Adding a multi-angle laser light scattering (MALLS) instrument to the multi-detector SEC setup allows the direct determination of molar masses as they elute from the SEC columns. The MALLS detector measures M_w and radius of gyration r_g for each SEC slice by an extrapolation to zero angle:

$$\frac{K \cdot c}{R_\theta} = \frac{1}{\overline{M}_w}\left[1 + \frac{4}{3}<r_g^2>_z k^2 \sin^2\left(\frac{\theta}{2}\right)\right] \tag{6}$$

where the optical constant K is defined by:

$$K = \frac{4\pi^2 n_0^2}{\lambda_0^4 N_L}\left(\frac{dn}{dc}\right)^2 . \tag{7}$$

and:

R_Θ ≡ reduced Rayleigh ratio (scattering intensity) under scattering angle Θ
$<r_g^2>_z$ ≡ radius of gyration
k ≡ wave number $(2\pi \cdot n_0/\lambda_0)$

Molecular weight determination requires the knowledge of the specific refractive index increment ν, which is dependent on copolymer composition. Copolymer refractive index increments ν_c can accurately be calculated for chemically monodisperse fractions, if comonomer weight fractions w_i and homopolymer ν_i values are known:

$$\nu_c = \sum_i w_i \cdot \nu_i \qquad (8)$$

Copolymer ν_c values are obtained by the multiple detection SEC method described above.

In the case of copolymers, light scattering investigations are even more complex, however. This is due to the fact, that copolymers may have a molecular compositional heterogeneity and that refractive increments at the scattering center may be different from the overall refractive index increment. Therefore, in general only apparent molar masses, M_{app}, for copolymers can be measured.

The relation between true copolymer molar masses M_{wc} and M_{app} is given by (8), (9):

$$\bar{M}_{app} = \bar{M}_{wc} + 2P\left(\frac{\nu_a - \nu_b}{\nu}\right) + Q\left(\frac{\nu_a - \nu_b}{\nu}\right)^2 \qquad (9)$$

where: $P = \Sigma_i M_i \, \delta w_i$
         $Q = \Sigma_i M_i \, \delta w_i^2$
and: $\nu = (dn/dc)$

P and Q depend on copolymer structure and physical and chemical polydispersity.

The dependence of apparent copolymer molar mass on solvent refractive index is shown in Figure 3. This figure shows the importance of the choice of the SEC eluent for a good agreement of M_{app} and true copolymer M_w. The solvent refractive index should be either significantly lower or higher than the specific refractive indices of both of the homopolymers of the comonomers forming the copolymer. An eluent with a refractive index between those of the comonomers will cause the apparent molar mass to diverge with respect to the true copolymer molar mass.

This general outline on light scattering treatment for copolymers seems to make copolymer characterization very complicated. If we apply this treatment to a chromatographic environment, things will get simpler as can be seen in the discussion of special cases below.

Case 1: Statistical Linear Copolymers:
SEC of random copolymers separates by molecular size. Therefore, there is

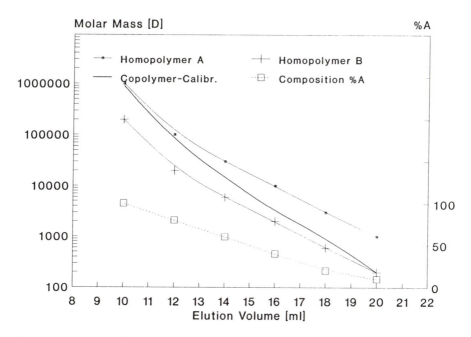

Figure 2. Construction of copolymer calibration curve based on homopolymer calibration curves and compositional information

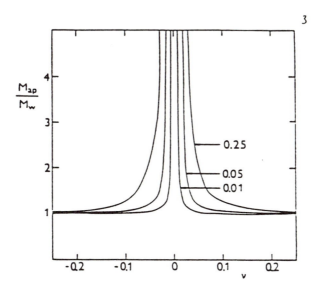

Figure 3. Dependence of M_{app}/M_w on copolymer refractive index increment ν and homopolymer refractive index increment difference $\nu_A-\nu_B$. This difference is also shown with the curves. Reproduced with permission from Ref. *(16)*. Copyright 1987 Elsevier.

a fractionation based on chain length also, despite the fact that the eluding copolymer fractions show a larger chain length variation than a homopolymer.

In the limiting case of statistical linear copolymers with narrow molar mass distribution a rigorous mathematical treatment shows (8):

$$P = 0$$

$$Q \approx M_o \, w(1-w)$$

Therefore:

$$M_{app} = M_{wc} \quad , \text{if } M > 10^4 \text{ D}$$

Case 2: block copolymers: Using the multi-detector SEC setup we can measure the variation of chemical composition with molar mass for a copolymer. Therefore, we are able to calculate copolymer refractive index increments according to eq. (8).

Most block copolymers are prepared by ionic polymerization techniques which in most cases will generate narrow molar mass distributions. But even if there is some physical heterogeneity, the chromatographic fractionation in the SEC column will result in pretty homogeneous fractions.

Calculation of P and Q factors for block copolymers with narrow MMD shows:

$$P = Q = 0$$

This means that the measured molar mass at any eluting fraction are equal to true copolymer M_{wc}.

On-line Viscometry. On-line viscometry opens a direct route to obtain copolymer molecular weights based on the universal calibration approach (3), (4), (10).

Universal calibration is based on the fact that SEC retention is dependent on the hydrodynamic radius of the solute under operating conditions. Application of Einstein's viscosity theory relates hydrodynamic volume to molar mass (11):

$$[\eta] \cdot M \propto V_h = f(V) \tag{10}$$

Consequently, at any given elution volume the following equation is valid:

$$[\eta]_1 \, M_1 = [\eta]_2 \, M_2 \tag{11}$$

Eq. (10) is independent on polymer architecture but does depend on temperature, solvent strength and chemical composition.

The on-line viscometer measures (copolymer) intrinsic viscosities directly.

Therefore, copolymer molar masses M_c can be calculated easily, if the polymer under investigation is homogeneous in polymer architecture and composition:

$$M_c = \frac{[\eta]_1 \, M_1}{[\eta]_c} \tag{12}$$

The measurement of $[\eta]_c$ by the viscometer and the subsequent calculation of M_c yields also the Mark-Houwink coefficients K_c and a_c of the copolymer/solvent system, which both are dependent on the chemical composition of the copolymer.

Additionally, copolymer M_n values can be calculated using Goldwasser's equation (1).

$$\bar{M}_n(abs) = \frac{3 \, \phi \, c_{inj} \cdot V_{inj}}{4 \pi V_s \sum_i \frac{(\eta_{sp})_i}{V_i}} \tag{13}$$

with:

c_{inj} $\equiv$ injected polymer concentration
V_{inj} $\equiv$ injection volume
Φ $\equiv$ Flory constant
V_s $\equiv$ volume of single slice

Experimental

All analyses were run in THF at ambient temperature on three PSS SDV 5 μm SEC columns (1000Å, 10^5Å and 10^6Å, 8 x 300 mm ea.) (PSS, Germany). The solvent was freshly distilled and degassed using an in-line membrane vacuum degasser (Erma, Japan). A Spectra Physics IsoChrom HPLC pump (Spectra Physics, USA) was used for eluent delivery and operated at 1.0 ml/min.

The effluent was divided into exactly equal volume streams to accommodate four detectors with minimum band broadening. The first flow path was equipped with the multi-angle laser light scattering detector (DAWN-F, Wyatt Technology, USA) connected serially to a Shodex SE 61 differential refractometer (Showa Denko, Japan).

The second flow path consisted of a Spectra Physics UV/VIS photometer plumbed in series with an on-line viscometer detector (Model 110, Viscotek, USA). For details of the experimental setup consult Figure 4.

Base calibration was done with PSS polystyrene standards (PSS, Germany). Conventional SEC calibration was performed using the corresponding homopolymer standards from the same vendor. In the case of star-shaped polystyrene and poly(*i*-propenyl naphthalene) universal calibration was applied (THF, 25°C: K = $1.496 \cdot 10^{-2}$ ml/g, a = 0.659 *(12)*).

Wyatt Technology's data capture and processing software (ASTRA 2.1) was

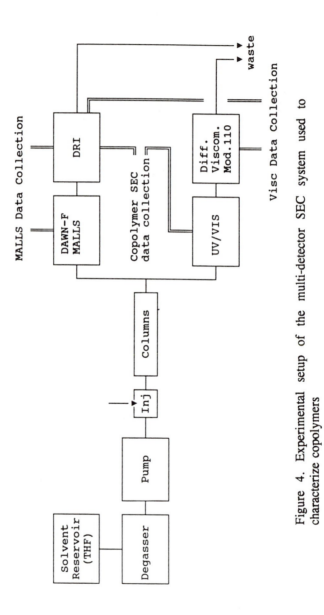

Figure 4. Experimental setup of the multi-detector SEC system used to characterize copolymers

used to calculate molecular weights. The data acquisition and processing of viscosities were done with Viscotek's UNICAL 3.11 software package.

PSS SEC software (V. 3.05) was used to extract molecular weight information from conventional GPC runs. SEC characterization of copolymers was done with PSS's COPO copolymer software package (V. 2.1), which allows the determination of the composition distribution and molecular weights of segmented copolymers simultaneously *(5)*, *(7)*, *(13)*.

Results and Discussion

Verification.

Validation of Composition Calculation Using Multiple Detection. The validity of copolymer composition determined by multiple detection was checked by the analysis of block copolymers consisting of styrene and methyl methacrylate segments (cf. Table I) as well as by random copolymers consisting of MMA and *t*-butyl methacrylate (cf. Table II) using NMR measurements, gravimetry and element analysis.

Table I: Composition of poly(styrene-*b*-MMA) determined by different methods

| sample | MMA content in % (w/w) determined by | | |
	comonomer ratio	^{1}H-NMR	COPO-SEC software
c13117	50	49	54
bc16117	52	--	52
bc7077	50*	24	21
bc16077	75*	41	37

* value too high due to incomplete conversion caused by termination

Multiple-detection SEC generally shows good agreement with direct methods independent on the type of copolymerization. Deviations can be explained by the different sensitivity of the bulk methods applied.

Validation of Copolymer Molar Masses. The precision of molar masses calculated for block copolymers were checked by on-line multi-angle laser light scattering. A 4-arm poly(styrene-*b*-butadiene) copolymer was analyzed by method 1 (multiple detection) revealing homogeneous butadiene content. SEC measures 65% (w/w) diene content, as compared to 67% in the comonomer feed. **Detector Calibration:** In order to obtain precise comonomer concentrations along the molar mass distribution the SEC detectors have to be calibrated. This was done with the corresponding homopolymers, which were injected at four or five concentrations in the range of 0.2 to 2 g/l.

Table II: Composition of poly(MMA-*co*-tBMA) determined by different methods

sample	bulk MMA content in % (w/w) determined by		
	GC	COPO-SEC	¹H-NMR
9.1	54.4*	64.4	63.8
9.2	62.7	62.4	60.1
9.3	56.1	53.0	53.1
9.4	52.4	52.3	52.3
9.5	47.4	46.8	47.1
9.6	43.0	43.6	
9.7	42.0	40.5	41.9
9.8	41.2	41.8	
9.9	40.2	41.1	41.0
9.10	39.4	39.2	
9.11	38.3	38.7	38.0

measurement by Dr. P. Mai, University of Mainz;   *) deviation due to low conversion

A comparison of these results with on-line MALLS demonstrate the validity of the assumptions mentioned above (eqs. (4) ff). Table III shows the molar masses for different parts of the copolymer distribution for determined by both methods.

Table III: molar masses measured by MALLS and calculated by SEC

	copolymer		4-arm star	
	SEC	MALLS	SEC	MALLS
$M_{n,c}$	151,000	154,000		
$M_{w,c}$	294,000	320,000		
$M_{p,c}$	93,000	92,000	383,000	380,000

Figure 5 shows a plot of peak area *vs.* injected concentration for poly(styrene) and poly(1,4-butadiene) in THF at 25°C using RI and UV detection at 260 nm according to eq. (1).

Detector response values for the homopolymers used in this investigation are given in Table IV.

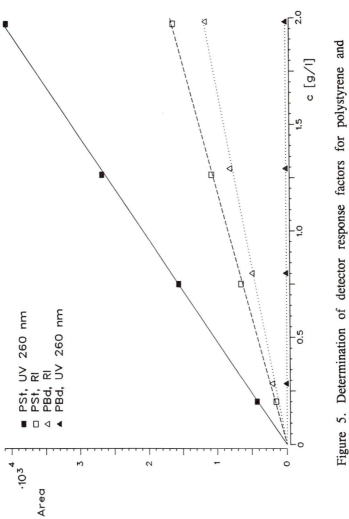

Figure 5. Determination of detector response factors for polystyrene and polybutadiene for RI and UV detection at 260 nm

Table IV: detector response factors for different comonomers (UV and RI detection)

monomer	f(UV260) [l/g]	Δf/f [%]	f(RI) [l/g]	Δf/f [%]
styrene	1981 ± 3.0	0.15	796.8 ± 1.8	0.22
butadiene	0.814 ± 0.11	1.32	593.7 ± 1.4	0.25
methyl methacrylate	6.26 ± 0.24	3.87	411.1 ± 0.7	0.14
n-butyl methacrylate	15.44 ± 0.17	1.10	422.4 ± 0.6	0.18
benzyl methacrylate	1193 ± 0.92	0.07	640.2 ± 1.5	0.24
i-propenyl naphthalene	35053 ± 18.8	0.05	1180 ± 0.49	0.04

The relative errors are small indicating a good precision in concentration determination. This has been confirmed by P. Mai *(14)*, where he could accurately determine the comonomer weight fractions of statistical MMA/*t*-BMA copolymers even at comonomer ratios of 99:1. Since these comonomers have nearly identical dn/dc values and similar adsorption coefficients, Mai's experiments indicate the scope of this technique. The trade-off of such precision is, however, a very high signal to noise ratio and a fairly high amount of detector calibration work.

Investigation of AB-type block copolymers. Block copolymers are very often prepared by ionic polymerization techniques using sequential addition of comonomers. If termination and transfer reactions are absent, block copolymers with narrow molecular weight can be synthesized.

General synthetic route (anionic polymerization as example):

$$I^{\ominus} + m A \xrightarrow{} I\text{-}A_m^{\ominus} \xrightarrow{+ n B} I\text{-}A_m\text{-}B_n^{\ominus}$$

Analysis of block copolymers by conventional techniques is relatively complex, however. In order to understand the polymerization reaction and to check the purity of the block copolymer, block yield and by-products have to be measured.

Conventional SEC only gives apparent molar mass distributions, which may help to identify side-reactions. Multiple-detection SEC, however, not only can measure comonomer composition distribution but can also be used to calculate molar masses. The combination of both information can be a very useful tool to understand and improve the copolymerization process.

This idea is illustrated in Figure 6, which shows the SEC molar mass and composition distribution for a styrene/MMA AB block copolymer being detected by RI and UV. The RI trace (– – –) records both styrene and MMA segments (according to their detector response factor), the UV photometer was tuned to a wavelength of 260 nm at which primarily the π-π^* transition of the styrene phenyl rings are detectable (———). In such cases even detector ratioing gives qualitative

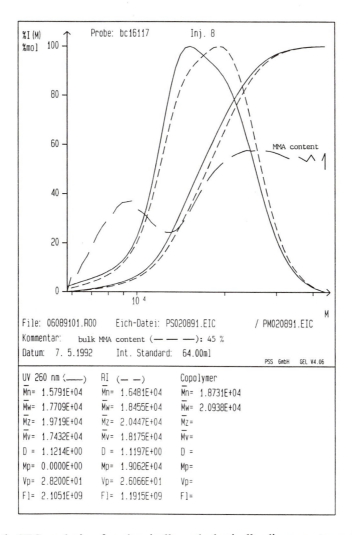

Figure 6. SEC analysis of a chemically and physically disperse styrene/MMA block copolymer; the dashed line (— — —) indicates the variation of the MMA content

compositional information. Quantitative comonomer content and copolymer molar masses can however be extracted only after detector calibration. If both monomers contribute at appreciable amounts to the response of all detectors, even qualitative interpretation of data is difficult with detector ratioing.

The molar mass distribution in Figure 6 is bimodal indicating physical and/or chemical heterogeneity. The variation of MMA content (———— ————) in the copolymer reveals the compositional heterogeneity of the styrene/MMA copolymer. The MMA content at the low molecular weight end is much lower than at the high molecular weight end.

We interpret the decrease in MMA content as a homo-polystyrene contamination caused by partial termination, when the second monomer (MMA) was added to the propagating polystyrene chains. Due to limited resolution, the MMA concentration does not reach zero as would be expected. Figure 6 also reports molar masses in polystyrene equivalents (first two columns) and the corresponding copolymer values (third column). As expected, copolymer molar masses are higher due to the incorporation of MMA into the calculation, which has higher segment density as compared to polystyrene.

If we compare the multi-detection data with on-line viscometry shown in Figure 7, the bimodality is more difficult to see and there is no indication of compositional changes at all. The molar masses, however, agree very well with data calculated by multiple detection and MALLS (see Table V for details).

Figure 8 shows a multi-modal molar mass distribution of an AB-BA type copolymer consisting of styrene and butadiene segments measured by multiple detection. Butadiene content is constant, apart from the low molar mass end, where it drops nearly to zero indicating a homo-polystyrene contamination.

This can be explained easily looking at the polymerization process: firstly styrene is polymerized to form segment A with subsequent addition of butadiene to yield a AB block structure. Finally the AB block is terminated using a bifunctional terminating agent X to give a AB-X-BA product with the same composition but double molar mass compared to the AB precursor. At the addition of the second monomer (butadiene) the polystyrene chain is partially terminated forming the homopolymer contaminant. On addition of the terminating agent a part of the AB block copolymer is deactivated by monofunctional species. We therefore see in the molar mass distribution with increasing molar mass: polystyrene (no butadiene content), AB type polymer (66% butadiene) and the final AB-BA product with the same butadiene content as the AB chain.

Molar masses calculated based on polystyrene calibration only are also given in Fig. 8 together with calculated copolymer molecular weights. Molar masses of the copolymer are lower as compared to their polystyrene equivalents because the butadiene segments have higher hydrodynamic volumes than polystyrenes of the same molar mass. A comparison of molar masses determined by different methods shows a good agreement of multiple detection data with MALLS (cf. Table V).

Star-shaped Block Copolymers. The synthetic route used to prepare a star-shaped styrene/butadiene AB copolymer is similar to the ones above. Star formation was done by a tetrafunctional terminating agent X. The arms consist of styrene/butadiene blocks:

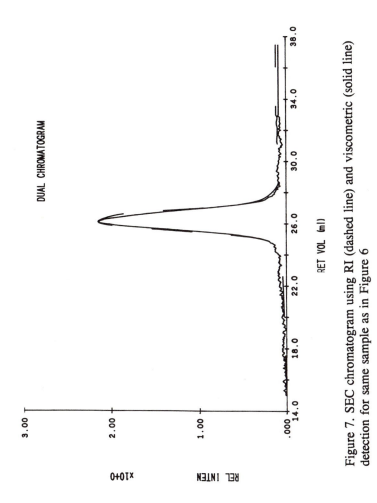

Figure 7. SEC chromatogram using RI (dashed line) and viscometric (solid line) detection for same sample as in Figure 6

Table V: Comparison of Copolymer Molar Mass Averages Determined by Different Methods

sample/type	lot #	M_n [·10³ D]			M_n	M_w [·10³ D]			composition in wt %
		SEC	MALLS	Visc	Goldw.	SEC	MALLS	Visc	
St-b-MMA	bc16117	18	18	18	18	20	20	20	52
St-b-MMA	bc13117	59	60	51	40	63	64	52	51
St-b-MMA	sm23028	100	99	62	86	110	107	69	23
St-b-MMA	sm12108	431	426	236	403	595	631	430	28
St-b-Bd	bdps31071	30	29	33	32	33	31	34	47
St-b-Bd	bdps27071	27	27	30	40	31	29	32	52
St-b-Bd	bdps12071	27	32	32	52	28	39	39	6
PS-b-Bd	bdps16071	17	14	16	22	18	16	17	66
St-b-Bd	BC5	115	118	125	71	135	153	137	55
St-b-Bd	BC4	167	186	92	67	214	234	135	67
(St-b-Bd)₄X	star BC	151	154			294	320		65
BMA-g-IPN (comb)	B8coBMA2	13	19	20	119	75	84	208	47
PS-g-IPN (comb)	Ac1coSty2	181	169	121	75	510	464	384	2.5
BzMA-co-MMA	FA100	36	73	54	38	52	93	72	12

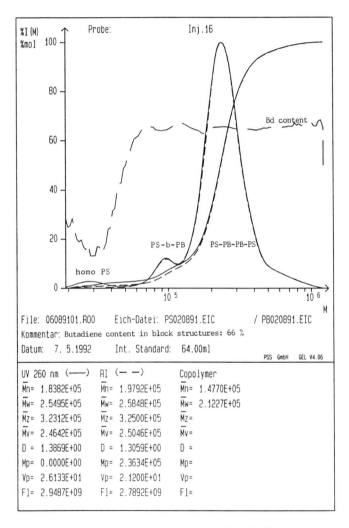

Figure 8. Molar mass and butadiene content distribution (———— ————) of a styrene/butadiene AB block copolymer as determined by multiple detection

$$I^{\ominus} \; + \; m \, A \; \xrightarrow{\hspace{2cm}} \; I\text{-}A_m^{\;\ominus} \; \xrightarrow{\; + \, n \, B \;} \; I\text{-}A_m\text{-}B_n^{\;\ominus}$$

$$4 \; I\text{-}A_m\text{-}B_n^{\;\ominus} \; + \; XR_4 \; \xrightarrow{\hspace{2cm}} \; \begin{array}{ccc} I\text{-}A_m\text{-}B_n & & B_n\text{-}A_m\text{-}I \\ & X & \\ I\text{-}A_m\text{-}B_n & & B_n\text{-}A_m\text{-}I \end{array}$$

The molar mass distribution as measured by multiple detector SEC is given in Figure 9. Three peaks and a high molar mass shoulder are visible. Despite the multi-modal MMD the butadiene content (65%) is constant throughout. Peak copolymer molar masses (M_{pc}) are denoted at the trace. As discussed earlier polystyrene equivalent molar masses deviate significantly from calculated copolymer values, which themselves agree very well with molar masses measured by on-line MALLS (cf. Table VI).

Table VI: measured (MALLS) and calculated molar masses of star PS-b-PB

	bulk copolymer		2-arm star		4-arm star	
	SEC	MALLS	SEC	MALLS	SEC	MALLS
$M_{n,c}$	151,000	154,000				
$M_{w,c}$	294,000	320,000				
$M_{p,c}$	93,000	92,000	174,000	170,000	383,000	380,000

Figure 10 shows the polystyrene (□) and polybutadiene (■) calibration curves and compares the calculated copolymer calibration curve (full line) with the one measured by on-line MALLS (▲). A good agreement is found between calculated and measured copolymer molecular weights apart from the very high and low molar masses, where polymer concentration is very low. The reason for deviations at low molar masses is probably the low signal to noise in the MALLS detector, at high molar masses indications for micro gel contaminations in the sample have been found *(15)*. This will lead to higher measured copolymer masses than being seen in concentration detectors only.

Comb-shaped Copolymers. In order to check the applicability of multiple detection, comb-shaped copolymers were also investigated. These consist of a backbone to which side chains with narrow molar mass distributions are attached in random positions by radical copolymerization of styrene and the ω-styryl *i*-propenyl naphthalene (IPN) macromonomer.

Figure 11 shows a molar mass distribution of a comb-shaped copolymer with a polystyrene backbone and IPN side-chains as measured by multiple detection. The IPN content (i.e. side chain density) is pretty low (2.5%), and its variation is minor also. No influence of side chain density is also seen in the linear Mark-

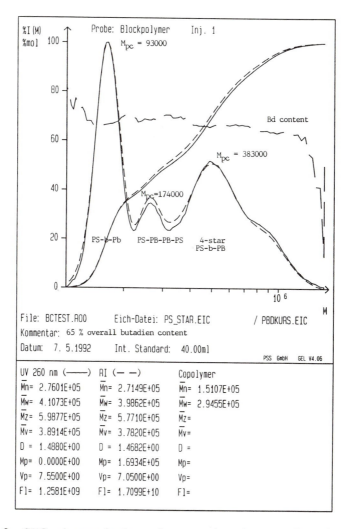

Figure 9. SEC characterization of a star-shaped styrene/butadiene block copolymer with multiple detection (———— ————, Bd content)

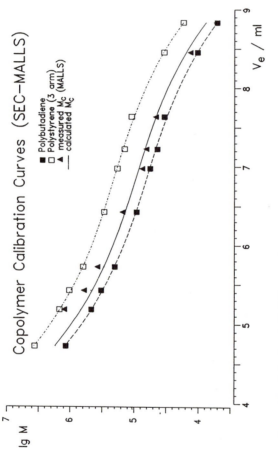

Figure 10. Calculated (full line) and measured (MALLS, ▲) calibration curve of star-shaped block copolymer from Figure 9. Black and open squares show individual homopolymer calibration curves

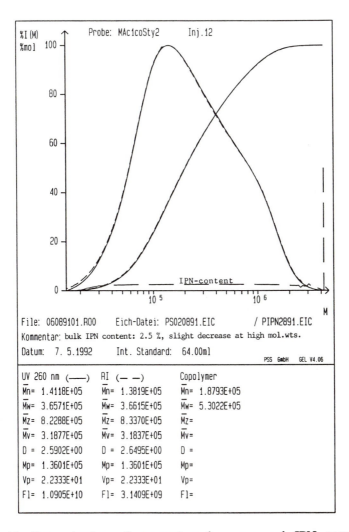

Figure 11. Determination of apparent molar mass and IPN composition distribution (———— ————) for comb-shaped styrene/IPN copolymer

Houwink plot constructed from measured intrinsic viscosities (by on-line viscometer) and measured molar masses (by MALLS). A strong change of side-chain frequency would be visible in a deviation from linearity in the Mark-Houwink plot (cf. Figure 12).

Careful investigation of composition information reveals, however, a decrease in side chain density with increasing molar mass. This is reasonable taking into account that incorporation of a macromonomer molecule gets the more difficult the larger the length of the main chain becomes. However, this change is too small to be visible in the double-logarithmic Mark-Houwink plot.

Molar masses determined by MALLS and multi-detection SEC agree quite well, those calculated based on universal calibration tend to be lower (cf. Table V). Due to its low IPN content treatment of this copolymer like a homopolymer seems applicable *(16)*. This indicates that multiple detection SEC can also give a good estimate on graft and comb-shaped copolymers with low comonomer concentrations.

Random Copolymers. A random copolymer synthesized from benzyl and methyl methacrylate was also studied to understand the limits of multiple detection. Figure 13 shows the results of SEC equipped with UV (at 260 nm) and RI detectors. The incorporation of benzyl methacrylate comonomer is pretty low (about 12%) with a trend to increase at higher molar masses. The composition agrees well with independent dn/dc and proton-NMR measurements (15% and 13% respectively).

A comparison of calculated or measured molar masses shows, however, a large disagreement. No absolute molar mass data are available at the moment to help to understand the validity of the measured data. Further investigations are currently carried out.

Conclusions

It has been shown that a SEC instrument equipped with multiple concentration and molar mass sensitive detectors is very useful to characterize copolymers. These measurements can be used efficiently to elucidate polymerization processes as was demonstrated for several block copolymers synthesized via anionic polymerization techniques.

Investigating chemical composition a very good agreement between multiple detection SEC and conventional bulk methods (NMR, GC, gravimetry) has been found independent of comonomer pairs and copolymer topologies. The SEC method has the additional advantage of being very simple and fast to carry out, but most important seems its potential to measure not only bulk composition but chemical composition distributions which had not been available previously without extensive solution/precipitation fractionations. Compositional together with copolymer molar mass information are a very useful tool to investigate polymerization processes and to understand copolymer properties better.

A good agreement of molar masses determined by copolymer SEC, on-line multi-angle laser light scattering and on-line viscometry was found for block copolymers, which have been studied extensively (cf. data given in Table V). The

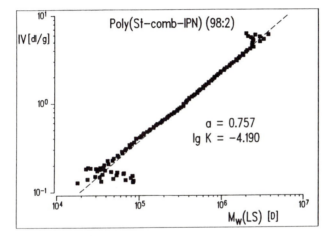

Figure 12. Linear dependence of measured intrinsic viscosity on molar mass (MALLS) for poly(styrene-<u>comb</u>-IPN) indicating a homogeneous side chain incorporation

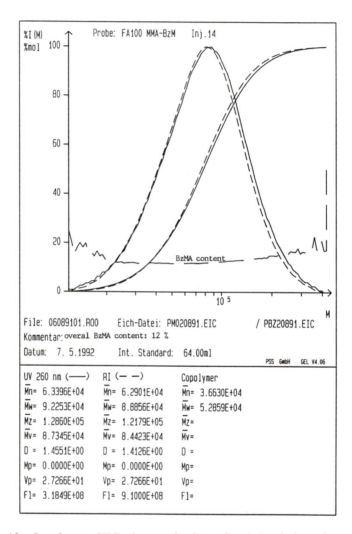

Figure 13. Copolymer SEC characterization of poly(methyl methacrylate-co-benzyl methacrylate). The dashed line (——— ———) represents the benzyl methacrylate incorporated into the random copolymer.

same holds true for a star-shaped block copolymer. The calculated copolymer calibration curve using multiple detection and that measured by MALLS agree very well. This seems to confirm our approach to calculated copolymer molar masses using homopolymer calibration curves and compositional information as long as heterocontacts are scarce as is the case in block and graft copolymers.

The analysis of comb-shaped polymers reveals a more complicated picture. At least for low side-chain densities multi-detector SEC and MALLS molar masses agree quite well with values calculated using universal calibration deviate more. Further studies are, however, needed for a better understanding of all parameters involved.

Copolymer M_n values as determined by Goldwasser's approach (equation (13)) using viscometric slice data promises great potential in its easy accessibility in a SEC run. The measurements presented here show sometimes very good agreement with other methods, sometimes large deviations for no obvious reason. It was generally found that the calculated M_n values depend very much on baseline settings and baseline noise. For polymers with very narrow molar mass distribution M_n's usually agreed well.

Integration of multiple concentration and molar mass sensitive detectors into one SEC unit has even greater potential when complex data handling and reduction problems will be solved. Then simple, flawless and accurate measurement of Mark-Houwink coefficients will be possible.

References:

1. Goldwasser, J.M.; *Proceedings International GPC-Symposium* **1989**, 150

2. Dondos, A.; Rempp, P.; Benoit, H.;   *Makromol. Chem.*, **1974**, *175*, 1659

3. Chang, F.S.C. et. al. *J. Chromatogr.*, **1971**, *55*, 67

4. Goldwasser, J.M.; Rudin, A. *J. Liq. Chromatogr.*, **1983**, *6*, 2433

5. Runyon, J.R.; Barnes, D.E.; Rudel, J.F.; Tung, L.H. *J. Appl. Polym. Sci.*, **1969**, *13*, 2359

6. Dondos, A.; Rempp, P.; Benoit, H. *Makromol. Chem.*, **1969**, *130*, 223

7. Kilz, P.; Johann, C. *Preprints 1ˢᵗ International Conference on Molecular Mass Characterization of Polymers*, Bradford, UK, 1989

8. Bushuk, W.; Benoît, H. *Can. J. Chem.* **1958**, *36*, 1616

9. Benoît, H.; Froelich, D. In *Light Scattering from Polymer Solutions;* Huglin, M.B., Ed., Academic Press, London, 1972; Chapter 11

10. Benoît, H.; Grubisic, Z.; Rempp, P.; Decker, D.; Zillox, J.-G. *J. Chim. Phys.* , **1966**, *63*, 1507

11. Tanford, C. *Physical Chemistry of Macromolecules*, Wiley, New York, 1961, p. 391

12. Engel, D. *PhD Thesis*, Mainz, 1980

13. Kilz, P. *GIT Fachz. Lab.* **1990**, *34*, 656

14. Mai, P. *PhD Thesis*, Mainz, 1990

15. Kilz, P.; Johann, C. *J. Appl. Polym. Sci., Symp.* **1991**, *48*, 111

16. Kratochvíl, P. In *Classical Light Scattering from Polymer Solutions;* Jenkins, A.D., Ed.; Polymer Science Library 5; Elsevier: Amsterdam, 1987; pp 203-215

RECEIVED October 21, 1992

Chapter 11

Size-Exclusion Chromatography and End-Group Analysis of Poly(methyl methacrylate)

Somnath S. Shetty and L. H. Garcia-Rubio

Chemical Engineering Department, University of South Florida, Tampa, FL 33620

This paper reports on the combined size exclusion chromatography and end-group analysis for poly methyl methacrylates produced at several temperatures and with different degrees of conversion. Spectroscopy analysis of the polymer, combined with molecular weight information, allows for the identification and quantification of the type and number of initiator fragments contained in the polymer. It is demonstrated that the residual initiator concentration may be obtained as an integral part of the analysis. The quantitative interpretation of end group-molecular weight analysis results in the identification of the termination mechanisms and in the estimation of the initiator efficiencies.

Initiation reactions constitute the forcing function for polymerization processes and are key to the understanding of the rate behavior of polymerization reactions and to the evolution of the molecular properties as a function of the reaction trajectory. In order to improve existing kinetic models, knowledge of the change in initiation efficiency as a function of conversion, and reliable estimates of the reaction rate constants are desired. In addition, to complete the kinetic model, it is important to consider an adequate gel effect model.

The deterrent factors in developing an improved polymerization model are as follows: the efficiencies of initiation reactions are unknown, the rate constants reported in the literature depend on the type of initiator used and there is a wide range of values reported in the literature for the same rate constants (1). Therefore, to complete   the understanding of the initiation kinetics and improve the estimates of the rate constants, it is desirable to obtain quantitative information on the type and number of initiator end groups present in the polymer. This vital piece of information is obtained with a combined end group analysis (EGA) from spectroscopy and molecular weight data from size exclusion chromatography (SEC).

0097–6156/93/0521–0149$08.25/0

In this paper, combined SEC and EGA for poly methyl methacrylates synthesized with benzoyl peroxide with different degrees of conversion is reported. The experiments were performed at two levels of temperature and initiator concentration. The outcome of the combined analysis is the identification of the type of initiator end-group in the polymer and quantification of the number of initiator fragments contained in the polymer. The unreacted initiator concentration is obtained as an integral part of the spectroscopy analysis. The interpretation of the quantitative end group-molecular weight analysis leads to the identification of the termination mechanisms and in the estimation of initiator efficiencies.

Benzoyl Peroxide was chosen for this study because it is one of the most common free radical initiators and there is a large body of literature concerning its chemistry and decomposition reactions *(2-4)*. In addition, it has been shown that the chromophoric groups resulting from the decomposition of BPO can be readily quantified. The BPO fragments attached to the polymer molecules as end groups can be expected to be products of the monomer and the benzoyl and phenyl radicals.

Experimental

A two level factorial design on temperature and initiator concentration was used to study the effects on the rate of polymerization and on the properties of the polymer produced *(5)*. The temperature levels were 60°C and 80°C, and the initiator concentration levels were 0.5% and 1% wt BPO. This design resulted in four distinct experiments. Replicate experiments at each of these conditions were performed to ensure that the experimental procedures were reproducible and to obtain adequate estimates of the experimental error.

Polymethyl methacrylates (PMMA) with benzoyl peroxide (BPO) end groups were synthesized via suspension polymerization in a 1 liter Kontes glass reactor. The temperature in the reactor was controlled within 1°C. The charge to the reactor consisted of a mixture of deionized water, poly vinyl alcohol as suspending agent, methyl methacrylate monomer (Aldrich Chemicals) and Benzoyl Peroxide (Fisher). The monomer with an initial purity of 99% was washed with NaOH solution and then dried with anhydrous magnesium sulphate. The monomer was then distilled under vacuum at about 30°C. BPO was purified by dissolving it in anhydrous methanol, recrystallizing twice and drying under vacuum at a low temperature.

The reaction was sampled at the desired intervals. At every sampling interval, two samples were withdrawn, one for conversion measurements, and the second for polymer characterization. The mixture sampled for conversion was dissolved in spectral-grade acetone for gas chromatography analysis. The polymer, from the sample for characterization, was precipitated twice from absolute methanol to eliminate the residual monomer.

Characterization

The reaction conversion was determined by measuring the weight of residual monomer in the sample collected, using gas chromatography and the weight of polymer produced by gravimetry. Intrinsic viscosities were determined in tetrahydrofuran (THF) at room temperature with a Cannon Ubbelhode capillary viscometer. The following equation was used for the estimation of the intrinsic viscosity.

$$(\eta_{sp}) = e + (\eta)C + K_H(\eta)^2 \qquad (1)$$

The experimental error e was estimated as an integral part of the measurements *(6)*.

End Group Analysis. UV spectroscopy was used to determine the number of initiator end-groups per polymer molecule. The decomposition products of BPO are many *(3,7)*, but only two active radicals, benzoyl radical (from primary and induced decomposition) and phenyl radicals (from secondary decomposition) are of significant importance in the initiation of polymer chains *(2,4)*. Prior NMR and UV studies have shown that the reactivity of the phenyl radicals with the monomer is very low compared to the benzoyl radicals *(2,4)*. In addition, spectroscopic analysis of the PMMA samples did not indicate the presence of phenyl groups, hence the presence of phenyl groups was considered to be insignificant. Therefore, a detailed UV study of the polymer spectrum permits the quantitative assessment of the initiation reactions. The information from the spectrum was used to determine the type and concentration of the BPO fragments that react with the MMA molecules. A comparison of the initiator decomposition rate with the number of initiator fragments attached to the polymer, leads to the calculation of the initiator efficiencies.

The first step in the polymer composition analysis is the identification and quantification of the UV spectra. This identification involves the selection of the model molecules for the BPO end-groups and the polymer backbone chain. The model molecules serve as the building blocks for generating the spectra which best represent the spectra of the polymer sample with unknown composition. Ethyl benzoate was used to represent the benzoate groups resulting from primary radical termination, and primary and induced decomposition of the initiator. BPO was used as the model molecule to represent the presence of residual BPO that may be present in the polymer. MMA monomer was used to represent the end-group formed by transfer to monomer reactions. Very high molecular weight polymethyl methacrylate initiated by azobisisobutyronitrile (AIBN) was used as the model molecule to represent the backbone of the polymer chain. The reason for choosing this model molecule is that AIBN is transparent to UV light in the region of interest and the relative concentration of AIBN is very small compared to the polymer itself.

The UV spectra for the polymers and model molecules in acetonitrile were measured in a Perkin Elmer 3840 photodiode array UV/VIS spectrophotometer equipped with a thermoelectric cell holder and a temperature controller with temperature programming capabilities. All measurements were taken at 25°C in a 1-cm path length cell. Up to seven concentrations plus replicates were used in order to obtain good estimates of the measurement errors (4). Special care was taken to ensure that the measurements were always within the linear range of the instrument. The spectrophotometer data was stored in a Perkin Elmer 7500 computer linked to a SUN 3/160 workstation for further processing with the interpretation software developed in-house.

Estimation of the polymer composition. In order to identify and quantify the concentration of the groups present in the polymer molecules, the standard assumptions of linearity of absorption with respect to concentration and additivity of the individual spectra of the components have been used (4). The above assumptions are reflected in the following equation.

$$A_j = \sum_{n=1}^{N} e_{nj} P_n lC \tag{2}$$

The linear behavior of the absorption with respect to concentration was statistically verified for all the samples analyzed. The extinction coefficient spectra for the polymer sample and the model molecules were estimated from their absorption spectra at different concentrations and are shown in Figures (1),(2), (3) and (4). Comparison of the extinction coefficient values in Figures (1),(3) and (4) displays the similarities between the spectra of the polymer sample and ethyl benzoate and BPO model molecules above 230 nm. The estimation of the extinction coefficients and compositions was done on a weight basis in order to minimize the propagation of experimental error (4). The calculated spectra are obtained from equation (2) applied to all the concentrations over all the measured wavelengths. The objective function used in the minimization process is given by (4),

$$Min: \sum_i \sum_j \frac{\phi_{ij}^2}{\sigma_{ij}^2} \tag{3}$$

where,

$$\phi_{ij} = A_{ij} - \sum_{n=1}^{N} e_{nj} P_n lC_i \tag{4}$$

$$\sigma_{ij}^2 = var(A_{ij}) + \sum_{n=1}^{N} (lC_i P_n)^2 \, var(e_{nj})$$

$$+ \sum_{n=1}^{N} (lP_n e_{nj}) \, var(C_i) \tag{5}$$

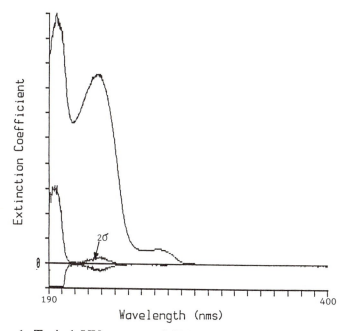

Figure 1. Typical UV spectra of PMMA synthesized with BPO. 95% confidence interval is indicated as 2σ.

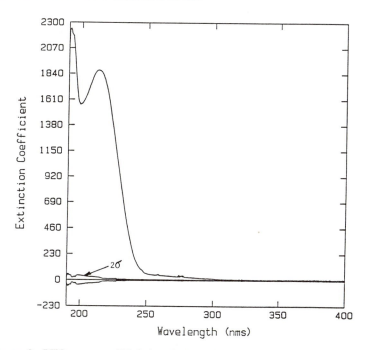

Figure 2. UV spectra of Poly(methyl methacrylate) model molecule. 95% confidence interval is indicated as 2σ.

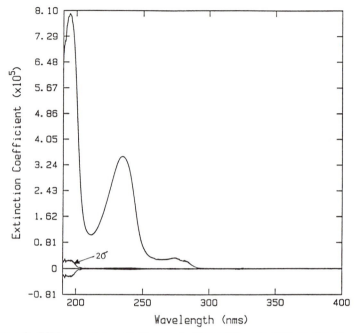

Figure 3. UV spectra of Benzoyl Peroxide model molecule. 95% confidence interval is indicated as 2σ.

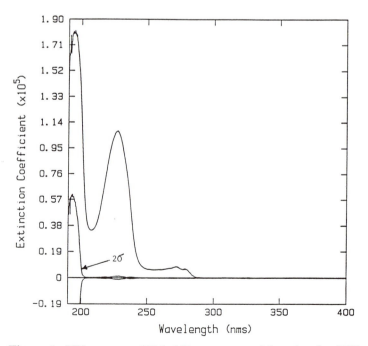

Figure 4. UV spectra of Ethyl Benzoate model molecule. 95% confidence interval is indicated as 2σ.

The solution to the non-linear equation (3) for the identified number of components yields the composition estimates. Figure (5) shows a typical comparison between the measured spectra for one of the polymer samples and the spectra calculated using model molecules. The difference between the measured and the calculated spectra is within the 95% confidence limits for the measured spectra as shown in the figure.

The wavelengths for the analysis were selected on the basis of the variances obtained for both the unknown samples and the model molecules. It was observed that for most polymer samples, the variances of the measurements increased sharply for absorbances measured below 230 nm. The reason for the increase in variance is the operation of the instrument in the region where the absorption is no longer linear. Therefore, in order to obtain reproducible concentration measurements with high signal to noise ratio and, at the same time, account for most of the spectral features, the analysis was limited to the range 230-400nm. The groups absorbing in this region are benzoates.

The end group analysis provides information on the grams of initiator fragments (end-groups) per gram of polymer produced. Combining this information with an independent measurement of the number average molecular weight, the number of end-groups per polymer chain can be readily calculated from:

$$\frac{Number \ of \ end\text{-}groups}{molecule} = \frac{grams \ of \ end\text{-}group}{grams \ of \ polymer} * \frac{M_n}{MW_{end\text{-}group}} \tag{6}$$

Molecular Weight Determination. The molecular weights were measured using size exclusion chromatography in THF at room temperature. The SEC experimental setup consisted of a solvent metering pump (Beckman Instruments 100A), a series of six Waters Associates microstyragel columns with pore sizes of 50,100,500, 10^3, 10^5 and 10^6 A°, a 1755 Biorad differential refractometer and an IBM computer for data acquisition and storage. The performance of the columns was tested by injecting a 0.5% solution of toluene in THF. The plate count was found to be within the manufacturer specifications.

The columns were calibrated with narrow polystyrene standards obtained from Polyscience and Pressure chemicals. The universal calibration equation was used to calculate the corresponding molecular weights for the poly methylmethacrylates. The polystyrene calibration curve is shown in Figure(6). The chromatograms were corrected for axial dispersion to minimize the measurement baises in the molecular weight averages, using Yau's technique (8). This technique implies that a linear molecular weight calibration and Tung's equation with gaussian spreading function are valid at each elution volume (V) (9). The number average and weight average molecular weight at each elution volume were determined from:

$$\overline{M_n}(V) = \frac{F(V)}{F\left(V + D_2\sigma^2\right)} e^{-\frac{1}{2}(D_2\sigma)^2} M_t(V) \tag{7}$$

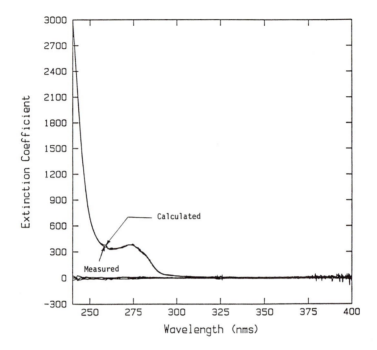

Figure 5. Comparison of the measured UV spectra and calculated spectra of a typical polymer sample.

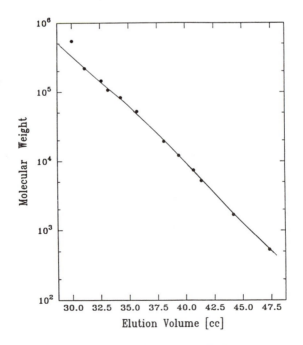

Figure 6. Calibration curve using narrow Polystyrene standards.

$$\overline{M_w}(V) = \frac{F(V - D_2\sigma^2)}{F(V)} e^{\frac{1}{2}(D_2\sigma)^2} M_t(V) \tag{8}$$

The number average and weight average molecular weight for the whole distribution of the polymer sample were calculated from:

$$\overline{Mw} = \frac{\sum F(V)\overline{Mw}(V)}{\sum F(V)}$$

$$\overline{Mn} = \frac{\sum F(V)}{\sum \dfrac{F(V)}{\overline{Mn}(V)}} \tag{9}$$

The value of σ in equations (7) and (8) is from the calibration curve for variance of the chromatograms obtained from the polystyrene standards (Figure (7)). Figure (8) shows a plot of the raw chromatogram and the chromatogram corrected for axial dispersion.

Results and Discussion

The conversion results at 60°C and 0.5 wt% BPO concentration (condition 1) and at 80°C and 1 wt% BPO concentration (condition 2), are shown in Figures (9) and (10). The rate of polymerization, as a function of conversion for both conditions is shown in Figures (11) and (12). As expected, the presence of a strong gel effect in both cases is clearly noticeable. The cumulative number average and weight average molecular weight obtained from size exclusion chromatography are shown in Figures (13) and (14) for both conditions. Notice the initial decrease in the molecular weights followed by an increase after approximately 30% conversion in agreement with the expected behavior from the rate behavior (Figures (11-12)). The viscosity average molecular weights obtained from the intrinsic viscosity measurements are shown in Figures (15) and (16) for the same set of conditions. Comparing Figures (13) and (15) and Figures (14) and (16) it is evident that the molecular weight averages from SEC are consistent with independently measured viscosity average molecular weights. Also, as expected, the viscosity average molecular weight values are in between the number average and weight average molecular weights. The experimental data at 60°C and 1 wt% BPO (condition 3) and 80°C and 0.5 wt% BPO (condition 4), in addition to conditions 1 and 2, are shown in Tables (I-VIII) in the appendix.

The number of end-groups per molecule can be calculated using equation (9) by combining the results from UV spectroscopy (Figures (17) and (18)) and the number average molecular weights obtained from SEC. The number of end-groups obtained from these measurements shed light into the termination mechanisms involved in the free radical polymerization of PMMA. If

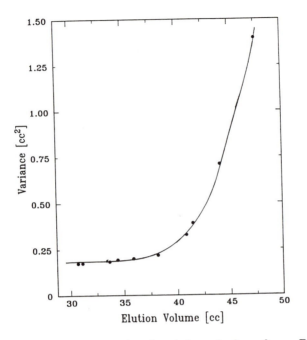

Figure 7. Calibration curve for band broadening, from Polystyrene standards.

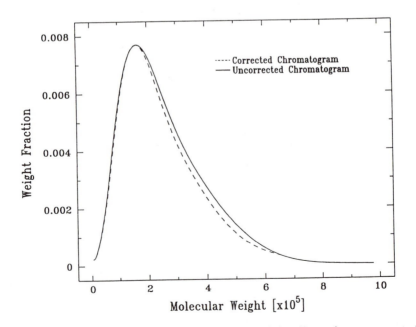

Figure 8. Typical chromatogram and axial dispersion corrected chromatogram.

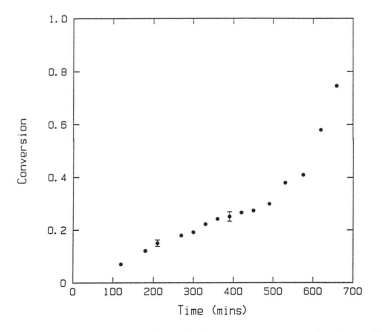

Figure 9. Conversion history for MMA suspension polymerization at 60°C and I_o = 0.5 wt% BPO.

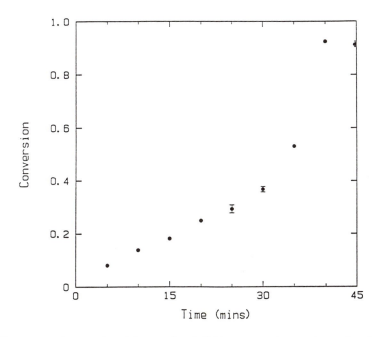

Figure 10. Conversion history for MMA suspension polymerization at 80°C and I_o = 1.0 wt% BPO.

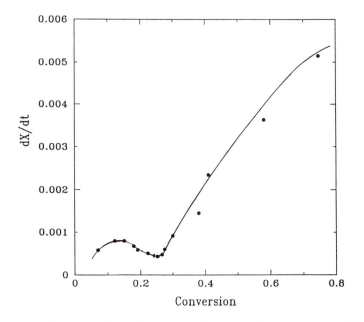

Figure 11. Rate of polymerization as a function of conversion at 60°C and I_0 = 0.5 wt% BPO.

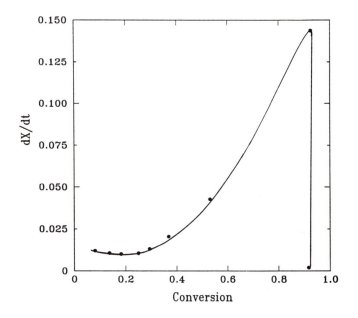

Figure 12. Rate of polymerization as a function of conversion at 80°C and I_0 = 1.0 wt% BPO.

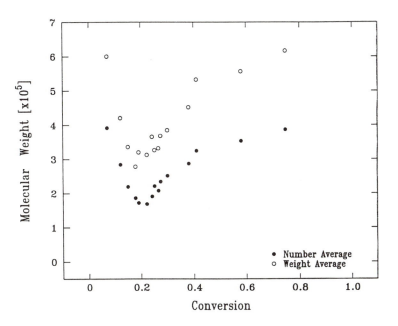

Figure 13. Cumulative number average and weight average molecular weight for MMA suspension polymerization at 60°C and $I_o = 0.5$ wt% BPO.

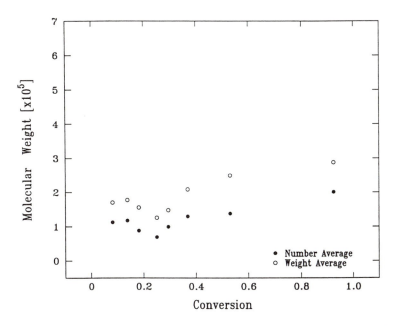

Figure 14. Cumulative number average and weight average molecular weight for MMA suspension polymerization at 80°C and $I_o = 1.0$ wt% BPO.

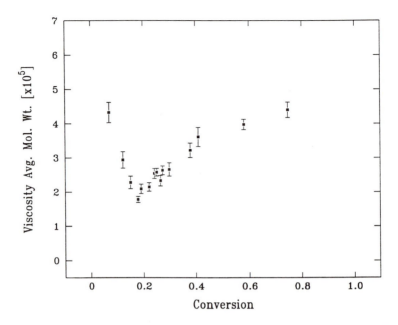

Figure 15. Viscosity average molecular weight versus conversion for MMA suspension polymerization at 60°C and I_o = 0.5 wt% BPO.

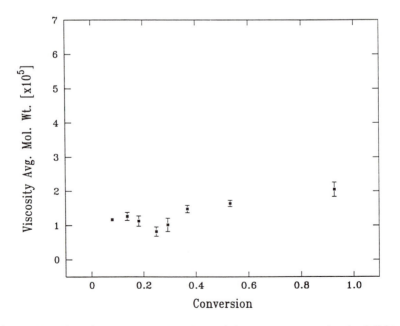

Figure 16. Viscosity average molecular weight versus conversion for MMA suspension polymerization at 80°C and I_o = 1.0 wt% BPO.

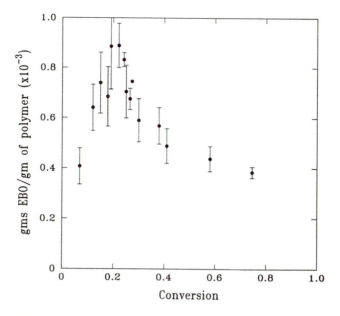

Figure 17. Grams of benzoate end-groups per gram of polymer as a function of conversion at 60°C and $I_o = 0.5$ wt% BPO.

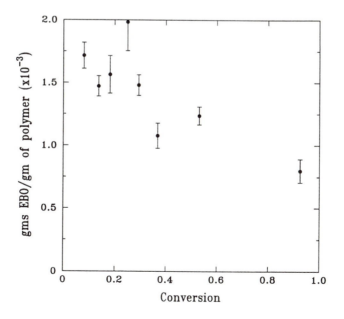

Figure 18. Grams of benzoate end-groups per gram of polymer as a function of conversion at 80°C and $I_o = 1.0$ wt% BPO.

termination is purely by combination, the number of end-groups should be 2; if termination is combination and, disproportionation or transfer to monomer, then the values should be between 1 and 2. However, if the termination is by disproportionation or transfer to monomer, then the values should center around 1. Figures (19),(20),(21) and (22) show the number of end-groups per polymer chain calculated at all four conditions. Comparison of Figures (20) and (22), i.e. at higher initiator concentration, suggests that primary radical termination may be of significance during the initial stages of the reaction. The number of end-groups and the molecular weight behavior also suggest that there is more than one mode of termination at these temperatures and initiator concentrations.

In a similar manner, the initiator efficiency may also be calculated from EGA. Figures (23),(24),(25) and (26) show the **ratio** Φ, defined as the grams of initiator end-groups bonded to the polymer per gram of initiator decomposed via first order thermal decomposition. If the initiator decomposes only through first order thermal decomposition, then Φ is a direct measure of the cummulative initiator efficiency. However, if other decomposition reactions, such as induced decomposition take place, Φ is not a direct measure of the initiator efficiency and could take values greater than 1. Figures 23-26 display the variety in behavior shown by MMA polymerizations. First of all, it is evident that first order thermal decomposition is not the only initiator decomposition mechanism in BPO/MMA polymerizations; secondly, the efficiency of initiation is not, in general, constant throughout the reaction trajectory; thirdly, the data suggests, in agreement with kinetic considerations, that the rate of initiator decomposition is a function of temperature and initial initiator concentration. It is seen in Figures (24-26) that Φ increases beyond 1, the limiting value, if the initiator decomposed only by first order thermal decomposition. This behavior suggests that more initiator radicals are being generated than by just first order thermal decomposition of the initiator, thus induced decomposition of the initiator may be present in MMA polymerizations. The ratio Φ at 60°C and 0.5 wt% BPO (Figure (23)) suggests that thermal decomposition dominates at lower temperature and lower initiator concentrations.

From the above results, it is seen that the combined size exclusion chromatography and spectroscopy experiments provide valuable information regarding the number of initiator end-groups per polymer molecule and thereby the termination mechanisms. Also, it is evident that important information regarding the efficiency of initiation is accessible through this type of measurements. It is expected that the use of UV spectrophotometers as SEC detectors will provide a detailed breakdown of the number of end-groups per polymer molecule as a function of the molecular size.

Nomenclature.

η	Intrinsic viscosity
η_{sp}	Specific viscosity
K_H	Huggins constant

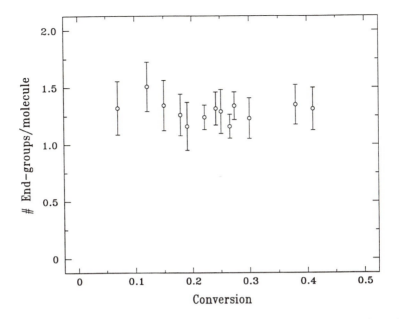

Figure 19. Number of end-groups per polymer molecule as a function of conversion at 60°C and I_o = 0.5 wt% BPO.

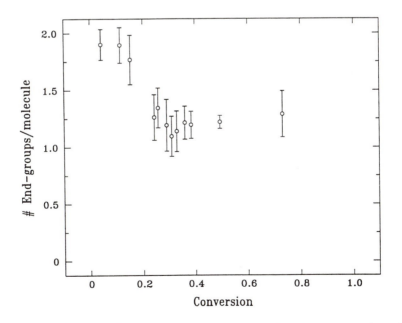

Figure 20. Number of end-groups per polymer molecule as a function of conversion at 60°C and I_o = 1.0 wt% BPO.

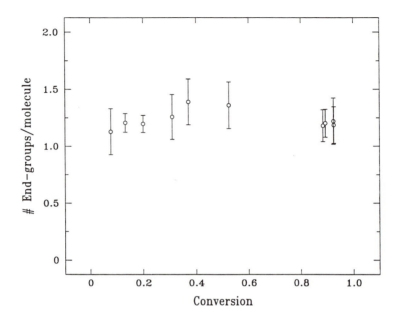

Figure 21. Number of end-groups per polymer molecule as a function of conversion at 80°C and I_o = 0.5 wt% BPO.

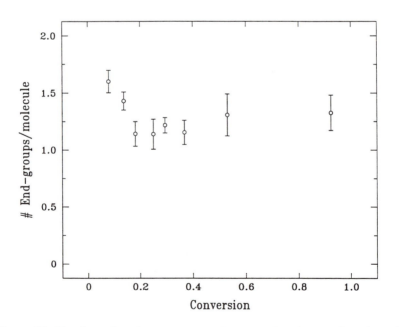

Figure 22. Number of end-groups per polymer molecule as a function of conversion at 80°C and I_o = 1.0 wt% BPO.

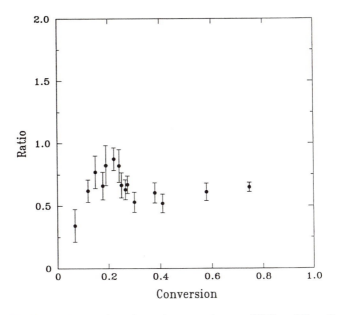

Figure 23. Ratio Φ as a function of conversion at 60°C and $I_o = 0.5$ wt% BPO.

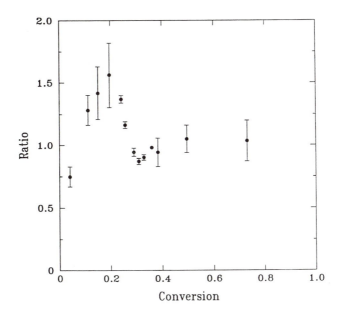

Figure 24. Ratio Φ as a function of conversion at 60°C and $I_o = 1.0$ wt% BPO.

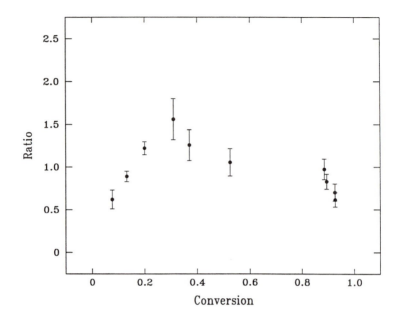

Figure 25. Ratio Φ as a function of conversion at 80°C and I_o = 0.5 wt% BPO.

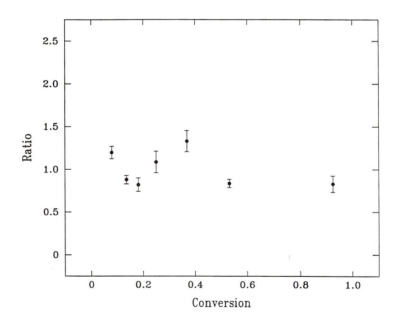

Figure 26. Ratio Φ as a function of conversion at 80°C and I_o = 1.0 wt% BPO.

A_j UV absorbance at wavelength j

e_{nj} Extinction coeffiicient for the nth component at the jth wavelength..

P_n Weight fraction of the nth component.

l Cell path length.

C Polymer concentration.

$\bar{M}n(V)$ Number average molecular weight as a function of the retention volume V.

$\bar{M}w(V)$ Weight average molecular weight as a function of the retention volume V.

$F(V)$ Experimental value of the chromatogram as a function of the retention volume V.

$M_t(V)$ Peak position molecular weight as a function of the retention volume V.

D_1, D_2 Constants related to the intercept and slope of the calibration curve.

σ   Polystyrene standards chromatogram standard deviation.

Literature Cited.

1. Polymer Handbook, 2nd edition; Brandrup,J. and Immergut,E.H.; John Wiley, NewYork, NY.
2. Moad, G.; Solomon, D. H.; Johns, S.; Willing, R. I., Macromolecules, **1982**,15, 1188.
3. Stickler, M.; Dumont, E., Makromol. Chem., **1984**, 187, 2663.
4. Garcia-Rubio, L. H.; Ro, N.; Patel, R.D., Macromolecules., **1984**, 17, 1998.
5. Shetty, S., PhD. Dissertation., University of South Florida, 1992.
6. Garcia-Rubio,L.H.; Talatinian,A.V. and MacGregor,J.F., Proc. Symp. on Quantitative Characterization of Plastics and Rubber, McMaster University, **1984**, 37
7. Tobolsky, A. V.; Rogers, C. E.; Brickman, R. D., J. Amer. Chem. Soc.,**1960**, 82, 1277.
8. Yau, W. W.; Stocklosa, H. J.; Bly, D. D., J. Appl. Poly. Sci., **1977**, 21, 1911.
9. Tung, L. H.; Moore, J. C.; Knight, G.W., J. Appl. Poly. Sci., **1966**, 10,

RECEIVED May 19, 1992

Appendix

Experimental data for BPO/MMA polymerizations.

Table I.   **Conversion vs Time Data at Temperature 60°C**

Initiator Concentration = 0.5% wt			Initiator Concentration = 1% wt		
Sample	Time (mins)	Conversion	Sample	Time (mins)	Conversion
2.1	120	0.07 +/-0.002	3.1	30	0.042
2.2	180	0.121 +/-0.003	3.2	60	0.113 +/-0.003
2.3	210	0.150 +/-0.012	3.3	90	0.152
2.4	270	0.179	3.4	120	0.197 +/-0.012
2.5	300	0.191 +/-0.006	3.5	140	0.242 +/-0.003
2.6	330	0.222	3.6	160	0.257 +/-0.001
2.7	360	0.242 +/-0.008	3.7	180	0.290 +/-0.005
2.8	390	0.251 +/-0.018	3.8	190	0.309 +/-0.002
2.9	420	0.266	3.9	200	0.329
2.10	450	0.274	3.10	210	0.360 +/-0.024
2.11	490	0.300 +/-0.002	3.11	220	0.384 +/-0.008
2.12	530	0.380 +/-0.003	3.12	240	0.495
2.13	575	0.410 +/-0.006	3.13	280	0.731
2.14	620	0.580			
2.15	660	0.746 +/-0.007			

Table II.   Conversion vs Time Data at Temperature 80°C

Initiator Concentration = 0.5% wt			Initiator Concentration = 1% wt		
Sample	Time (mins)	Conversion	Sample	Time (mins)	Conversion
4.1	10	0.078	5.1	5	0.081 + /-0.002
4.2	15	0.133 + /-0.008	5.2	10	0.138 + /-0.005
4.3	20	0.20	5.3	15	0.182 + /-0.007
4.4	30	0.31 + /-0.002	5.4	20	0.250 + /-0.002
4.5	40	0.371 + /-0.011	5.5	25	0.294 + /-0.015
4.6	45	0.525 + /-0.007	5.6	30	0.368 + /-0.010
4.7	50	0.885 + /-0.013	5.7	35	0.531 + /-0.004
4.8	60	0.894 + /-0.013	5.8	40	0.925 + /-0.008
4.9	70	0.917	5.9	45	0.914 + /-0.017
4.10	75	0.925			
4.11	80	0.919			
4.12	85	0.926			

Table III.   Viscosity Average Molecular Weight Data at Temperature 60°C

Initiator Concentration = 0.5% wt			Initiator Concentration = 1% wt		
Sample	Conv.	Viscosity Avg. Molecular Wt.	Sample	Conv.	Viscosity Avg. Molecular Wt.
2.1	0.07	432847+/-29595	3.1	0.042	377539+/-21110
2.2	0.121	293947+/-24129	3.2	0.113	256324+/-18090
2.3	0.150	228065+/-18471	3.3	0.152	225238+/-12560
2.4	0.179	178406+/-8714	3.4	0.197	182221+/-7580
2.5	0.191	209396+/-13632	3.5	0.242	169738+/-11675
2.6	0.222	215169+/-12480	3.6	0.257	166576+/-13217
2.7	0.242	254402+/-14754	3.7	0.290	200170+/-11773
2.8	0.251	258120+/-11097	3.8	0.309	195277+/-17687
2.9	0.266	233372+/-15482	3.9	0.329	209267+/-7917
2.10	0.274	263643+/-12735	3.10	0.360	204516+/-20530
2.11	0.300	266082+/-19405	3.11	0.384	209931+/-11832
2.12	0.380	322223+/-21023	3.12	0.495	226949+/-16796
2.13	0.410	361222+/-28054	3.13	0.731	300067+/-16771
2.14	0.580	397998+/-14871			
2.15	0.746	440589+/-22500			

Table IV.   Viscosity Average Molecular Weight Data at Temperature 80°C

Initiator Concentration = 0.5% wt			Initiator Concentration = 1% wt		
Sample	Conv.	Viscosity Avg. Molecular Wt.	Sample	Conv.	Viscosity Avg. Molecular Wt.
4.1	0.078	172799 + /-16868	5.1	0.081	117057 + /-5786
4.2	0.133	142927 + /-16415	5.2	0.138	126415 + /-11846
4.3	0.20	127719 + /-13884	5.3	0.182	112739 + /-15182
4.4	0.31	91624 + /-9646	5.4	0.250	82266 + /-13600
4.5	0.371	127204 + /-13367	5.5	0.294	102062 + /-19581
4.6	0.525	187166 + /-17493	5.6	0.368	148064 + /-10810
4.7	0.885	263534 + /-21269	5.7	0.531	164366 + /-9203
4.8	0.894	282192 + /-17348	5.8	0.925	205210 + /-21050
4.9	0.917	270976 + /-14531			
4.10	0.925	289706 + /-12800			
4.11	0.919	298809 + /-18606			
4.12	0.926	295693 + /-19981			

Table V. Number Average and Weight Average Molecular Weight Data at Temperature 80°C

Initiator Concentration = 1% wt			
Sample	Conversion	Number Average Molecular Wt.	Weight Average Molecular Wt.
5.1	0.081	112840	170388
5.2	0.138	117583	177550
5.3	0.182	88250	155609
5.4	0.250	69487	125789
5.5	0.294	99457	147745
5.6	0.368	129638	208542
5.7	0.531	138115	249532
5.8	0.925	200915	287999
5.9	0.914		

**Table VI. Number Average and Weight Average Molecular Weight
Data at Temperature 80°C**

Initiator Concentration = 0.5% wt			
Sample	Conversion	Number Average Molecular Wt.	Weight Average Molecular Wt.
4.1	0.078	148034	242695
4.2	0.133	126842	207894
4.3	0.20	104326	176067
4.4	0.31	89401	133252
4.5	0.371	111312	189893
4.6	0.525	164362	262138
4.7	0.885	235381	388464
4.8	0.894	239469	396059
4.9	0.917	244854	417721
4.10	0.925	241612	405183
4.11	0.919	246107	419381
4.12	0.926	239283	408416

Table VII. Number Average and Weight Average Molecular Weight
Data at Temperature 60°C

Initiator Concentration = 0.5% wt			
Sample	Conversion	Number Average Molecular Wt.	Weight Average Molecular Wt.
2.1	0.07	391939	600343
2.2	0.121	284261	420706
2.3	0.150	219532	335884
2.4	0.179	186412	278326
2.5	0.191	172027	320079
2.6	0.222	168735	312782
2.7	0.242	191408	365589
2.8	0.251	221494	326520
2.9	0.266	208281	332183
2.10	0.274	234703	368484
2.11	0.300	251846	385324
2.12	0.380	287273	452217
2.13	0.410	324905	532666
2.14	0.580	353647	556640
2.15	0.746	387568	617072

Table VIII. Number Average and Weight Average Molecular Weight Data at Temperature = 60°C

Initiator Concentration = 1.0% wt			
Sample	Conversion	Number Average Molecular Wt.	Weight Average Molecular Wt.
3.1	0.042	307605	556842
3.2	0.113	241351	359773
3.3	0.152	182879	322361
3.4	0.197	148295	255735
3.5	0.242	138954	232200
3.6	0.257	162481	257909
3.7	0.290	178261	289050
3.8	0.309	179356	289702
3.9	0.329	182141	293702
3.10	0.360	186053	292494
3.11	0.384	194216	305336
3.12	0.495	209348	310251
3.13	0.731	286506	455545

SIZE-EXCLUSION CHROMATOGRAPHY: VISCOMETRY DETECTION

Chapter 12

A Strategy for Interpreting Multidetector Size-Exclusion Chromatography Data I

Development of a Systematic Approach

Thomas H. Mourey[1] and Stephen T. Balke[2]

[1]Analytical Technology Division, Research Laboratories, Eastman Kodak Company, Rochester, NY 14650–2136
[2]Department of Chemical Engineering and Applied Chemistry, University of Toronto, Toronto, Ontario M5S 1A4, Canada

A systematic approach for diagnosing and overcoming problems in multidetector SEC analysis is presented. The system examined consists of differential refractive index (DRI), low-angle laser light-scattering (LALLS) and differential viscometry (DV) detectors. An ultraviolet (UV) detector was also used to monitor an internal standard used as a flow rate marker. The strategy underlying this approach is to sequentially select the values to be calculated from the data so as to be able to isolate sources of error. A broad standard with known molecular weight averages is injected. Data from the DRI and conventional calibration curve are used first. Once problems have been resolved and results are satisfactory, each molecular weight detector is examined in turn. Finally, data from both the DRI and the molecular weight detectors is used to obtain interdetector volumes. It is shown that by using this approach, separate calculations of the molecular weight distribution for the broad standard using conventional narrow standard calibration, LALLS and DV (with universal calibration) provide identical results. For the analysis of broad samples on this moderately high-resolution system, resolution correction proved unnecessary and axial dispersion did not affect the interpretation used. However, it is shown that if narrow molecular weight distribution polymers are to be analyzed, resolution correction must be considered. An example utilizing the "Method of Molecular Weight Averages" is shown.

The interpretation of multidetector SEC data is predictedly more complex than the interpretation of conventional SEC data obtained from a single, concentration-sensitive detector. Molecular-weight-sensitive SEC detectors exemplify this point; processing data from DV and LALLS detectors requires instrument calibration constants, sample constants, concentrations, injection

0097–6156/93/0521–0180$06.00/0

volumes, and flow rates. In many instances, it is difficult to identify which parameters are incorrect even when obviously erroneous results are obtained. This paper shows the development of an approach to systematically diagnose problems and to establish the values of parameters which provide accurate molecular weight distributions. The strategy underlying this approach is to sequentially select the values to be calculated from the data so as to be able to isolate sources of error.

In the following sections, first the multidetector system used to develop the approach is described and the effect of system parameters on the observed molecular weight distributions is shown. The simple steps composing the approach are then listed, and then each step is discussed in turn along with the results obtained.

Experimental

A schematic of the multidetector SEC is shown in Figure 1. Four 10 μm particle-diameter, 7.8 mm i.d. x 300 mm Plgel columns (Polymer Laboratories, Amherst, MA) are connected in series. The eluate passes through a Spectroflow Model 757 UV-visible detector and is then split approximately equally to a Model 100 DV (Viscotek Corp., Porter, TX) and a KMX-6 LALLS photometer (LDC Analytical, Riviera Beach, FL) by regulating the backpressure on the DV branch. A Waters model 410 (Waters, Division of Millipore Corp., Milford, MA) differential refractive index (DRI) detector is placed in series after the LALLS photometer. The columns, DV and DRI are thermostated to 30.0°C. The eluent is uninhibited tetrahydrofuran (THF, J. T. Baker, Inc.), pumped at a nominal flow rate of 1.00 mL/min. Concentrations of narrow polystyrene standards (Polymer Laboratories) between 7,800,000 MW and 580 MW range from 0.06-2.5 mg/mL, for high-to-low molecular weights, respectively. NBS 706 broad-molecular-weight distribution polystyrene is dissolved at a concentration of 1.5 mg/mL. Injection volumes are 100 μL. To each sample solution is added 0.01% 1-chloro-2,4-dinitrobenzene (CDNB, Kodak Laboratory and Research Products, Rochester, NY) as a flow rate marker, which is monitored at 350 nm by the UV detector.

Effects of Data Processing Parameters

Motivation for the development of a systematic approach is the diversity of parameters which must be specified for analysis, combined with the similarity of the effects of different parameters on the detector outputs. A few examples illustrate these points. Curve (a) of Figure 2 is the molecular weight distribution of NBS 706, calculated from the DV and DRI detector responses and a universal calibration curve. Curve (b) of Figure 2 is obtained by reprocessing the data with a flow rate that is 1% lower than the true value. The distribution is shifted to higher molecular weight but the shape remains unchanged. If flow rate alone were the only source of this shift, it would be relatively easy to identify; however, an error in either the sample concentration (Figure 3) or in the value of the differential pressure transducer calibration factor [this is an instrument

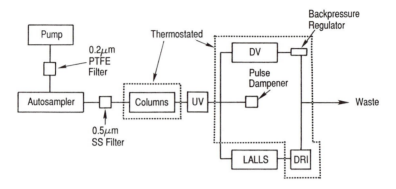

Figure 1. Schematic of multidetector SEC.

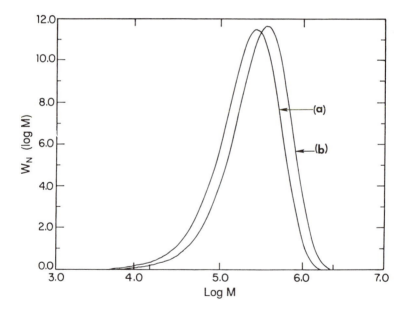

Figure 2.   Effect of an error in flow rate on the molecular weight distribution of NBS 706 calculated from DV detector and universal calibration; a) true distibution, b) flow rate 1% lower than true value.

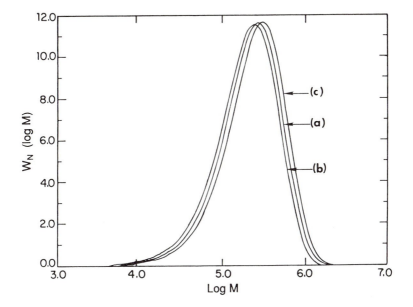

Figure 3. Effect of an error in the sample concentration (set 10% low) on the molecular weight distribution of NBS 706; a) true distribution, b) distribution from DV and universal calibration for sample concentration 10% lower than true value, c) distribution from LALLS for concentration 10% low.

calibration factor sometimes referred to by the instrument vendor as the "DPT sensitivity factor", that converts the millivolt output of the differential pressure transducer to Pascals] also shifts the molecular weight distribution without affecting curve shape (Figure 4). Similar shifts are also caused by errors in injection volume, DV inlet pressure, or in any of the constants used for the calculation of LALLS scattering intensities.

Changes in the shape of the molecular weight distribution are produced by errors in the interdetector volume (the effective volume between a molecular-weight-sensitive detector and the DRI detector, Figure 5) or improper resolution correction (Figure 6). There is no accompanying shift in the molecular weight distribution on the log M axis, which distinquishes these errors from the shifts shown in Figures 2-4; however, we cannot distinquish between the effects of the value for interdetector volume and improper resolution correction on the shape of the molecular weight distribution.

The Systematic Approach

Figure 7 is a schematic of the systematic approach which helps to identify and eliminate uncertainty in data processing parameters that cause shifts and changes in molecular weight distributions. There are only four simple steps involved. Following each step, sources of error are diagnosed and removed before proceeding to the next step. The steps are summarized as follows:

Step 1. Inject a broad standard (NBS 706 polystyrene) of known molecular weight averages and utilize the DRI output and the conventional calibration curve to obtain M_n, M_w, and M_z.

Step 2. Calculate the total intrinsic viscosity for NBS 706 from the DV output,

$$[\overline{\eta}] = \frac{1}{m}\int_0^\infty \eta_{sp}(v)dv \tag{1}$$

where $\eta_{sp}(v)$ is the specific viscosity as a function of retention volume v and m is the total mass injected. A similar equation is used to calculate the whole polymer M_w from the LALLS output,

$$\overline{M}_w = \frac{1}{mK}\int_0^\infty R_\theta(v)dv \tag{2}$$

where K is the LALLS optical constant and $R_\theta(v)$ is the excess Rayleigh scattering as a function of retention volume v. Quantities calculated using Equations 1 and 2 are independent of chromatographic resolution and require only that the entire sample elute from the column.

Step 3. Determine the interdetector volume between the DV and DRI detectors by a search for the value of interdetector volume which successfully superimposes the plot of local intrinsic viscosity versus retention volume for NBS

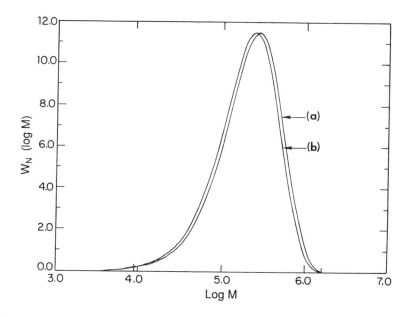

Figure 4.   Effect of an error in the differential pressure transducer calibration factor on the molecular weight distribution of NBS 706; a) true distribution, b) distribution from DV and universal calibration for calibration factor 10% lower than true value.

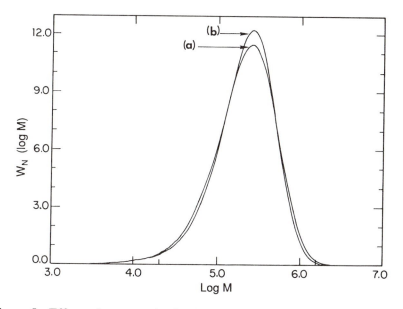

Figure 5.   Effect of an error in the interdetector volume on the molecular weight distribution of NBS 706 calculated from DV; a) true distribution, b) interdetector volume set 50 μL lower than true value.

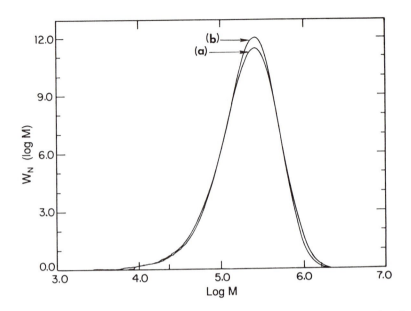

Figure 6. Effect of an improper axial dispersion correction on the molecular weight distribution of NBS 706 calculated from DV; a) true distribution, b) improper resolution correction.

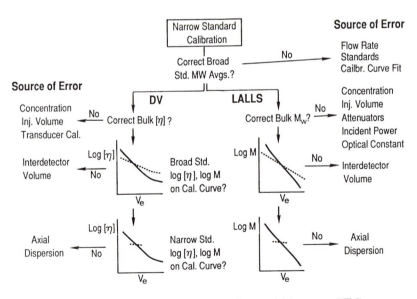

Figure 7. Systematic approach for multidetector SEC.

706 on the plot of total intrinsic viscosity of each narrow standard (from DV) versus retention volume(*1-3*). The local intrinsic viscosities, $[\bar{\eta}]_i$, are calculated from $\eta_{sp,i}$ (from the DV detector) and the concentration at each retention volume, c_i, obtained from the DRI,

$$[\bar{\eta}]_i = \frac{\eta_{sp,i}}{c_i} \tag{3}$$

[Note: $\eta(v)$ refers to the specific viscosity as a function of v, whereas $\eta_{sp,i}$ refers to a particular value of specific viscosity at a particular retention volume i.]
Local weight-average molecular weights, $M_{w,i}$, are calculated from the LALLS and DRI detectors, assuming that the second virial coefficient A_2 is negligible,

$$\overline{M}_{w,i} = \frac{R_{\theta,i}}{Kc_i} \tag{4}$$

and the interdetector volume between the LALLS and DRI detectors is determined by superimposing log M versus retention volume curves.

Step 4. Assess and correct for axial dispersion effects, if necessary. For analysis of a broad molecular weight distribution polymer using high-resolution columns and room-temperature operation, it is likely that an assessment will show that there is no need for resolution correction. However, for analysis of a narrow molecular weight distribution sample, this will not be the case. Then, a variety of options are available. Some details on assessment and resolution correction are provided here while others may be found in Part II of this work. In the following sections, each step is discussed in turn along with the results obtained.

Discussion

Step 1. Step 1 utilizes the concentration-sensitive (DRI) detector only. A broad standard of known $\overline{M}_n$, $\overline{M}_w$, and $\overline{M}_z$ is required. The mean values for NBS 706 polystyrene given in Table 1 were obtained from conventional SEC over a 2 year period and have been confirmed by LALLS. Comparison with literature values for NBS 706 obtained by SEC and other techniques was presented in a previous paper(*4*). If these values are not obtained when NBS 706 is analyzed by conventional, narrow-standard calibration, then three common sources of error are examined.

1. Incorrect flow rate. All calculated molecular weight averages of the NBS 706 will be systematically higher or lower than the correct values. For best results, the flow rate for the broad standard run should be correct to 0.01% with respect to the flow rate measured at the time of narrow standard calibration. Methods that utilize internal flow markers have been used for several years to correct flow rate fluctuations (i.e., reference 5). The correct flow rate, Q_c, can be calculated from the retention volume t_{obs} of the flow rate marker (CDNB) in the broad standard, measured from the UV detector response at 350 nm,

$$Q_c = Q_r \left(\frac{t_r}{t_{obs}} \right) \tag{5}$$

where Q_r and t_r are the flow rate and retention time of CDNB at the time of narrow calibration standard. The value of Q_r is obtained during the elution of the first narrow calibration standard by timing the collection of 10 mL of eluent in a calibrated volumetric flask. Equation (5) corrects flow rates for run-to-run variations and normally provides the required accuracy for acceptable results. It does not compensate for flow rate variations *within* a run. Experience has shown that variations within a run are less significant than long-term flow rate fluctuations. Other methods of obtaining flow rates accurate to 0.01% are also acceptable. [Note: The UV detector placement before a parallel configuration, as shown in Figure 1, cannot correct for flow fluctuations in the individual branches of the parallel detector configuration. Such fluctuations, caused by a change in the flow split, are minimized by careful regulation of temperature and backpressures. Serial detector arrangements may be preferred if this flow split cannot be carefully regulated.]

Table 1. NBS 706 Polystyrene

	$\overline{M}_n$	$\overline{M}_w$	$\overline{M}_z$
Nar. Std. SEC (this study)	123,000	276,000	435,000
1σ[a]	5,700	4,600	8,900
95% conf. limit[a]	700	600	1,100
	$[\overline{\eta}]$	$\overline{M}_w$	
correct	0.094[b]	275,000[c]	
DV or LALLS, equations 1 or 2	0.094 ± 0.02[d]	274,000 ± 3,000[d]	

[a]Number of replicates N = 256
[b]Ubbelohde glass capillary
[c]static LALLS
[d]95% confidence, N = 9

2. Incorrect molecular weight values for narrow calibration standards. Some or all of the calculated molecular weight averages of NBS 706 may be incorrect. The use of a LALLS detector provides one means to check the accuracy of the molecular weight values of narrow standards quoted by the vendor.

3. A poor fit to narrow standard calibration data. One or all of the calculated molecular weight averages of NBS 706 may be incorrect. Methods to evaluate fits to calibration data using molecular-weight-sensitive detectors and the analysis of residuals have been discussed previously(4).

Other sources of error in conventional, concentration-sensitive SEC detection and conventional calibration can also contribute(6). The significant point is that the quality of data obtained from multidetector SEC data is highly dependent upon our ability to perform conventional SEC (i.e., SEC with only a DRI). Thus, we proceed to Step 2 after we are satisfied with the values of M_n, M_w, and M_z provided by our conventional SEC analysis.

Step 2. Step 2 optimizes each molecular-weight-sensitive detector. The correct whole polymer values of $[\bar{\eta}]$ and M_w are given in Table 1. Two potential sources of error in calculating these quantities from SEC utilizing Equations (1) and (2) are incorrect sample concentration or injection volume. These can result from weighing and dilution errors or inaccurate injection volumes. These errors can be more easily distinguished from other sources of shifts in molecular weight distributions because opposite changes are observed in LALLS and DV distributions (Figure 3). Any of the parameters used to calculate $\eta_{sp,i}$ or $M_{w,i}$ from the molecular-weight-sensitive detectors will also cause shifts in the molecular weight distribution. For DV, these parameters include the inlet pressure and the transducer calibration factor (if required). For LALLS, they include attenuator factors, incident laser power, and the optical constant K. In some instances the source of these errors can be difficult to isolate and correct, requiring careful evaluation of the performance and operation of each molecular-weight-sensitive detector. We proceed to Step 3 only after values of $[\bar{\eta}]$ and M_w are in satisfactory agreement with the correct values of Table 1.

Step 3. Step 3 combines the responses of the concentration-sensitive detector (DRI) and a molecular-weight-sensitive detector (DV or LALLS). Examples of the effect of interdetector volume on $[\bar{\eta}]_i$ and the molecular weight distribution calculated from DV and universal calibration are given in Figures 8 and 9. Studies with LALLS have shown that numerical searches for the interdetector volume using narrow standards(7) circumvent some problems associated with differences between the geometric and effective interdetector volumes(8). We have shown previously, however, that a numerical search for the interdetector volume using a single broad standard gives results as good as or better than the best available alternatives(1). For broad standards, there is sufficient experimental and theoretical justification to assume that the local values calculated from equations 3 and 4 are not significantly affected by axial dispersion(9,10). The optimum value provides the best superposition of $[\bar{\eta}]_i$ and $M_{w,i}$ on the intrinsic viscosity and molecular weight calibration curves (Figure 10, example for DV).

In some instances the correct molecular weight distribution may not be obtained from DV using universal calibration after obtaining the interdetector volume from Step 3. This may be caused by a poor fit to the universal

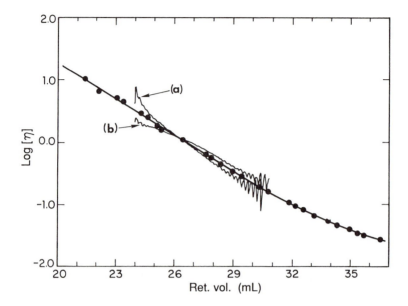

Figure 8. Effect of the value of interdetector volume on local intrinsic viscosities of NBS 706 measured by DV; a) interdetector volume = -0.10 mL, b) 0.16 mL.

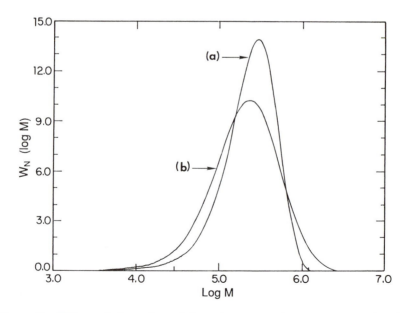

Figure 9. Effect of the value of interdetector volume on the molecular weight distribution of NBS 706; a) interdetector volume = -0.10 mL, b) 0.16 mL

calibration data or by incorrect values of the whole polymer $[\bar{\eta}]$ of some narrow standards. In both cases, examination of plots of residuals for log $[\eta]$ or log $M[\eta]$ versus retention volume narrow standard calibration curves will identify the problem(4).

By the conclusion of Step 3, we obtained comparable molecular weight distributions and molecular weight averages for NBS 706 from the three different methods: conventional narrow standard calibration and use of the DRI alone, DV with universal calibration and the DRI, and LALLS combined with the DRI (Figure 11). The values and precision of molecular weight averages calculated from LALLS and DV were reported in a previous paper(4).

Another criterion for considering whether or not the multidetector system is operating satisfactorily is to examine the constants of the empirical Mark-Houwink relationship,

$$[\eta] = KM^a \tag{6}$$

where the coefficient K and exponent *a* are specific to a given polymer-solvent system. From DV detection we obtain $[\eta] = 0.000147M^{0.706}$ for NBS 706. The exponent of this relationship is similar to the generally accepted value of 0.70 (*11*).

Thus, the first three steps of the approach provide some confidence that broad molecular weight distribution polymers can be adequately analyzed. However, even for this moderately high-resolution SEC system, analysis of narrow molecular weight distributions is not satisfactory. This leads us to the final step of the approach.

Step 4. Step 4 is the assessment and correction of resolution. The results obtained in Steps 1 through 3 demonstrated that for analysis of broad-molecular-weight- distribution samples, resolution was high and correction was unnecessary. However, it was observed that the interdetector volume obtained from Step 3 does not accurately superimpose $M_{w,i}$ and $[\bar{\eta}]_i$ of narrow standards on their respective log M and log $[\eta]$ calibration curves (example for DV given in Figure 12). Molecular weight distributions calculated for these narrow standards from DV are broader than the correct distribution (Figure 13), whereas LALLS molecular weight distributions are narrower. The Mark-Houwink exponent for the example shown in Figures 12 and 13 provided an unrealistic value of 0.367 when calculated from DV detection. Assuming that a Gaussian function is an adequate descriptor of band spreading in SEC, then the first requirement of resolution correction is determination of the standard deviation, σ, of this function at different retention volumes. One method of accomplishing this is to search for the value of σ which, when used together with the interdetector volume obtained for NBS 706 from Step 3, will enable the superposition mentioned above.

The equation utilized to correct the local, uncorrected intrinsic viscosity as a function of retention volume, $[\bar{\eta}](v,uc)$, (these values are calculated from Equation 3) was(*12*)

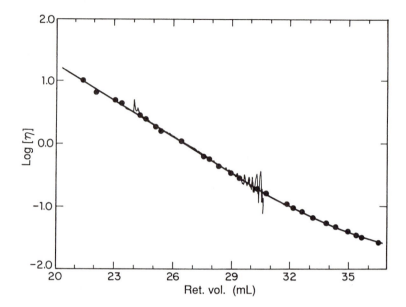

Figure 10. Superposition of local intrinsic viscosities of NBS 706 measured by DV on the narrow standard intrinsic viscosity calibration curve (symbols). The value of the interdetector volume is 0.029 ± 0.009 mL (95% confidence, 9 replicates).

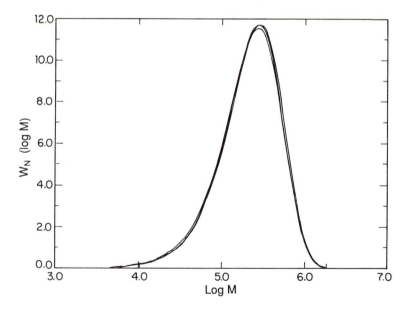

Figure 11. Molecular weight distributions of NBS 706 obtained by narrow-standard calibration, DV and universal calibration, and LALLS detection.

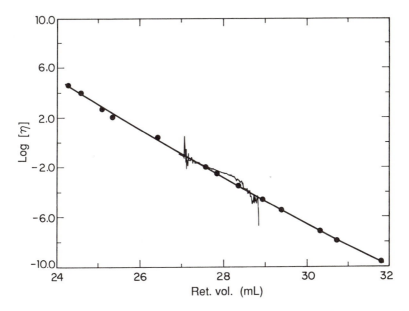

Figure 12. Local intrinsic viscosities of narrow-molecular-weight-distribution polystyrene 127,000 using the interdetector volume = 0.029 mL obtained from step 3, compared to the narrow standard intrinsic viscosity calibration curve (symbols).

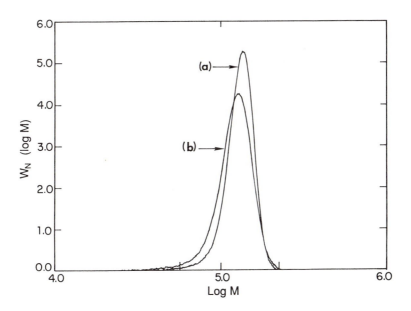

Figure 13. Molecular weight distribution of polystyrene 127,000 calculated by a) narrow standard calibration, b) DV and universal calibration using the interdetector volume = 0.029 mL obtained from step 3, without resolution correction.

$$[\overline{\eta}](v) = [\overline{\eta}](v,uc) \times \frac{F(v)e^{-(D_{2\eta}(v)\sigma(v)^2)/2}}{F(v-D_{2\eta}(v)\sigma(v)^2)} \qquad (7)$$

where $F(v)$ is the DV response at retention volume v, the denominator is the DV response at a retention volume less than retention volume v, and $D_{2\eta}$ is proportional to the slope of the narrow standard intrinsic viscosity calibration curve,

$$D_{2\eta} = -2.303 \times \frac{d\log[\eta]}{dv} \qquad (8)$$

Analogous equations are used for LALLS,

$$\overline{M}_w(v) = \overline{M}_w(v,uc) \times \frac{F(v)e^{-(D_2(v)\sigma(v)^2)/2}}{F(v-D_2(v)\sigma(v)^2)} \qquad (9)$$

and

$$D_2 = -2.303 \times \frac{d\log M}{dv} \qquad (10)$$

Similar searches for σ have been applied to LALLS by previous investigators(13). Values of σ obtained from the search increase with increasing molecular weight (Figure 14); however they are all smaller than $\sigma = 0.30$ measured from the CDNB peak. This implies that the search provides "effective" values of σ and that some compensation for axial dispersion is also obtained from the interdetector volume search in Step 3. The magnitude of the axial dispersion correction is often evaluated from the quantities $D_2\sigma^2$ and $D_{2\eta}\sigma^2$ used in Equations 7 and 9. It is significant to note that axial dispersion corrections for $[\overline{\eta}](v)$ become quite small at low molecular weights because both σ and $D_{2\eta}$ decrease with decreasing molecular weight. In contrast, the value of D_2 is more nearly constant for the entire log M calibration curve.

Once the variation of σ and $D_{2\eta}$ with retention volume have been determined, a variety of methods of correcting resolution are available(14,15). One of these methods, the "Method of Molecular Weight Averages"(16) involves correcting the whole polymer molecular weight averages for Gaussian dispersion according to the following equation:

$$\overline{M}_q = \overline{M}_q(uc)e^{(3-2q)D_2^2\sigma^2/2} \qquad (11)$$

where $q = 1, 2,$ or 3 for $\overline{M}_n$, $\overline{M}_w$, and $\overline{M}_z$. Results of applying Equation (11) for narrow polystyrene standards are given in Table 2. Values for polydispersities

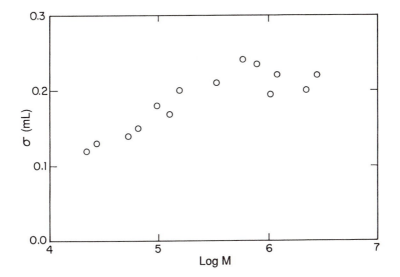

Figure 14. Effective values of σ obtained from step 4 for narrow polystyrene standards.

($\overline{M}_w/\overline{M}_n$) obtained by either DV or LALLS detection are similar to values obtained from narrow standard calibration, and Mark-Houwink exponents a are close to the expected values of 0.70.

Table 2. Narrow Polystyrene Standards

M	Vendor M_w/M_n	Nar. Std. M_w/M_n	DV M_w/M_n	a	LALLS M_w/M_n
2,730,000	1.04	1.06	1.04	0.766	1.06
2,170,000	1.06	1.06	1.04	0.676	1.07
1,117,000	1.06	1.02	1.05	0.734	1.02
1,030,000	1.05	1.05	1.05	0.692	1.03
775,000	1.04	1.06	1.06	0.746	1.04
570,000	1.05	1.06	1.04	0.703	1.06
336,000	1.03	1.03	1.03	0.711	1.03
155,000	1.03	1.04	1.03	0.691	1.03
127,000	1.05	1.02	1.02	0.719	1.04
96,000	1.04	1.04	1.06	0.712	1.05
66,000	1.03	1.03	1.05	0.691	1.05
52,000	1.03	1.02	1.02	0.711	1.07
26,700	1.03	1.03	1.03	0.702	1.02
22,000	1.03	1.02	1.04	0.671	1.03

The results of Step 4 can be practically applied to unknown polymers with narrow molecular size distributions (e.g., $M_w/M_n < 1.3$ for this column set) by correcting $[\overline{\eta}](v)$ and $M_w(v)$ using the value of σ for a narrow standard with a retention volume range similar to the unknown polymer. In most cases a polymer of the same chemical structure as the narrow unknown is available or can be synthesized. Values for D_2 and $D_{2\eta}$ required to correct for imperfect resolution of the unknown can be obtained from the LALLS and DV responses of this broad polymer.

Conclusions

A systematic approach to diagnosing and overcoming problems in multidetector SEC analysis was developed. Following injection of a broad-molecular-weight-distribution standard, four simple steps are required to implement the approach: (1) calculation of molecular weight averages using the DRI and conventional calibration curve; (2) calculation of $[\bar{\eta}]$ and $\bar{M}_w$ from the DV and LALLS detectors, respectively; (3) determination of the interdetector volume by numerical search for the most effective value; (4) resolution assessment and correction. At each step, sources of error are diagnosed and the situation corrected before progressing to the next step.

Good results were obtained by applying the first three steps to a broad-molecular-weight-distribution polystyrene standard. Step 4 contains many options and is often unnecessary. One case in which it is necessary is the analysis of narrow-molecular-weight-distribution polymers. Implementation of Step 4 to such polymers by utilizing a search for the standard deviation of the spreading function, followed by application of the "Method of Molecular Weight Averages," is shown.

Acknowledgements

We are very grateful to Ms. Sally Miller for generating the data necessary for this study and for testing the worthiness of the systematic approach.

Literature Cited

1.   Mourey, T. H.; Miller, S. M. *J. Liq. Chromatogr.* **1990**, *13*, 693
2.   Balke, S. T.; Cheung, P.; Lew, R.; Mourey, T. H. *J. Liq. Chromatogr.* **1990** *13*, 2929
3.   Cheung, P.; Balke, S. T.; Mourey, T. H. *J. Liq. Chromatogr.*, in press
4.   Mourey, T. H.; Miller, S. M.; Balke, S. T. *J. Liq. Chromatogr.* **1990**, *13*, 435
5.   Patel, G. N. *J. Appl. Polym. Sci.* **1974**, *18*, 3537
6.   Mori, S. In *Steric Exclusion Liquid Chromatography of Polymers*; Janča, J., Ed.; Chromatographic Science Series; Marcel Dekker Inc.: New York, NY, 1984, Vol. 25; Chap. 4.
7.   Lecacheux, D.; Lesec, J. *J. Liq. Chromatogr.* **1982**, *5*, 2227
8.   Bruessau, R. In *Liquid Chromatography of Polymers and Related Materials - II*; Cazes, J.; Delamare, X., Eds.; Chromatographic Science Series; Marcel Dekker, Inc.: New York, NY, 1980, Vol. 13; pp 73-93.
9.   Styring, M. G.; Armonas, J. E.; Hamielec, A. E. *J. Liq. Chromatogr.* **1987**, *10*, 783
10. Hamielec, A. E.; Meyer, H. In *Developments in Polymer Characterisation - 5*; Dawkins, J. V. Ed.; Elsevier Applied Science: London, 1986; pp. 95-130
11. Benoit, H.; Grubisic, Z.; Rempp, R. *J. Polym. Sci. Part B*, **1967**, *5*, 753
12. Hamielec, A. E. *Pure & Appl. Chem.* **1982**, *54*, 293
13. Zhi-Duan He, Xian-Chi Zhang, Rong-Shi Cheng, *J. Liq. Chromatogr.* **1982**, *5*, 1209

14.   Balke, S. T. *Quantitative Column Liquid Chromatography -A Survey of Chemometric Methods*, J. Chromatogr. Library; Elsevier: Amsterdam, 1984, Vol. 29; p. 81-82

15.   Yau, W. W.; Kirkland, J. J.; Bly, D. D. *Modern Size-Exclusion Liquid Chromatography*; John Wiley & Sons: New York, NY, 1979, p. 106

16.   Hamielec, A. E.; Ray, W. H. *J. Appl. Polym. Sci.* **1969**, *13*, 1319

RECEIVED April 20, 1992

Chapter 13

A Strategy for Interpreting Multidetector Size-Exclusion Chromatography Data II

Applications in Plastic Waste Recovery

Stephen T. Balke[1], Ruengsak Thitiratsakul[1], Raymond Lew[1], Paul Cheung[1], and Thomas H. Mourey[2]

[1]Department of Chemical Engineering and Applied Chemistry, University of Toronto, Toronto, Ontario M5S 1A4, Canada
[2]Analytical Technology Division, Research Laboratories, Eastman Kodak Company, Rochester, NY 14650–2136

The systematic approach for multidetector size-exclusion chromatography (SEC) interpretation is applied to high temperature SEC analysis of recycled plastic waste. The SEC was equipped with a differential viscometer (DV) in addition to the usual differential refractometer (DRI). A clean blend of polyethylene with polypropylene originating from diaper manufacturing was analyzed before and after processing to increase its molecular weight. The systematic approach was successfully used to set the needed operating conditions and parameters. Complicating factors influencing this application were axial dispersion effects, low molecular weight interference of the stabilizer peak with the main polymer peak and column degradation. Assessment of axial dispersion, experimental correction of the interference problem and careful attention to flow rate corrected chromatograms of narrow standards were found to be the most satisfactory responses to these respective factors.

In Part I, the development of a strategy for interpreting multidetector size-exclusion chromatography (SEC) data was shown (1). The reason for development of the approach is the diversity of parameters which must be specified for multidetector SEC analysis, combined with the similarity of the effects of different parameters on the detector outputs. In Part I, polystyrene standards were analyzed using a room temperature SEC with tetrahydrofuran as the mobile phase. This paper shows our initial attempts to utilize the strategy for a considerably less ideal system: a high temperature SEC and analysis of recycled plastic waste.

Theory

The equations essential to application of the systematic approach have already been provided in Part I. The emphasis here will be on application of these

0097–6156/93/0521–0199$06.25/0

fundamental equations to complex polymers (e.g. blends of different branched and linear homopolymers).

Presentation of Specific Viscosity Chromatograms. For the DRI detector, chromatograms are generally presented as normalized height versus retention volume (2). The normalized heights are obtained by dividing each detector response height by the area under the raw chromatogram. Each area increment on the DRI chromatogram then represents a weight fraction. However, for the chromatogram of specific viscosity versus retention volume obtained from the DV, a similar normalizing approach does not provide curves with useful physical significance . One alternative is evident from the equation used to obtain the whole polymer intrinsic viscosity:

$$\overline{[\eta]} = \int_0^\infty \left(\frac{\eta_{sp}(v)}{m} \right) dv \tag{1}$$

where $\overline{[\eta]}$ is the whole polymer intrinsic viscosity, $\eta_{sp}(v)$ is the local specific viscosity as a function of retention volume, v, (i.e. the total volume of eluent which has passed through the columns up to that time) and m is the mass injected. [Note: $\eta_{sp}(v)$ refers to the specific viscosity as a function of v while $\eta_{sp,i}$ refers to a particular value of specific viscosity at a particular retention volume, v_i.]

From this equation, it can be seen that presentation of the DV chromatogram as a plot of specific viscosity per unit mass injected ($\eta_{sp,i}/m$) versus retention volume provides a chromatogram with area equal to the whole polymer intrinsic viscosity. Also, this method of presentation removes mass injected as a source of differences in the chromatograms.

Calibration. In applying the systematic approach to this system, two methods are used to obtain calibration curves. The first (used here for polystyrene only) is to determine it by the conventional method of plotting the peak molecular weights of narrow molecular weight distribution standards versus their peak retention volumes. The second is to obtain it from a combination of the intrinsic viscosity detector response and the refractometer response. To accomplish this, the intrinsic viscosity at each retention volume (i.e. the local intrinsic viscosity, $[\eta]_i$) must be obtained from the DV by using:

$$[\eta]_i = \frac{\eta_{sp,i}}{c_i} \tag{2}$$

To obtain the c_i corresponding to each $\eta_{sp,i}$ the interdetector volume must be known [see next section].   Then the calibration curve for the sample analyzed is obtained from the usual ordinate of the universal calibration curve at a particular v_i, J_i (2), (i.e. the product of intrinsic viscosity and molecular weight) divided by the measured intrinsic viscosity from the DV detector at that v_i.

$$M_i = \frac{J_i}{[\eta]_i} \tag{3}$$

A calibration curve specific to each complex polymer sample was obtained from Equations 2 and 3. Since different branching, composition and molecular weight can combine together to provide the same molecular size in solution, there can be a distribution of molecular weights at any retention volume. Then, Equation 3 provides the local number average molecular weight of the variety of molecular weights present at each retention volume (*3*).

The Intrinsic Viscosity Calibration Curve. The "intrinsic viscosity calibration curve" is a plot of logarithm of the intrinsic viscosity versus retention volume. Two methods were used to determine this curve. The first involved correlating the whole polymer intrinsic viscosity of each narrow polystyrene standard (obtained by applying Equation (1) to the DV output for the narrow standards) with retention volume. The second was correlating the individual local intrinsic viscosity values obtained from applying Equation (2) at each retention volume for a broad standard with retention volume.

In this study, polyethylene-polypropylene blends were analyzed before and after processing. The processing was expected to increase their molecular weight by increasing the degree of long chain branching. An increase in branching was expected to increase both molecular weight and molecular size in solution. The increase in size may not be as much as if the molecules remained linear; however, some increase is expected. If the molecules remain linear, then the new intrinsic viscosity calibration curve and molecular weight calibration curve derived from Equation (3) will superimpose on the data obtained from the sample before reaction, with most of the data points now at the high molecular weight end. If branching occurs, then the new intrinsic viscosity calibration curve will lie below those for the unreacted sample. The molecular weight calibration curve derived from Equation (3) will then show that a higher molecular weight is now exiting at the same retention volume. Observation of the change in both the molecular weight calibration curve and the intrinsic viscosity calibration curve provides the initial basic information required to carry out more detailed branching analyses.

Interdetector Volume. The two above mentioned calibration curve determination methods for the intrinsic viscosity calibration curve, when applied to a broad molecular weight distribution polystyrene standard, formed the basis for determining interdetector volume (*1,4-5*). When the correct interdetector volume is specified, the two intrinsic viscosity calibration curves will superimpose if a broad molecular weight distribution standard is used with Equation (2)(*1,4-5*). Thus, in this paper a numerical search is used to find the interdetector volume which will accomplish this superposition.

The interdetector volume determined is really an "effective one" in that no attempt is made to correct for different cell sizes amongst the detectors and some effects of axial dispersion can influence results. However, in a recent comparison with other methods, this method appears of high practical utility (*6*).

Axial Dispersion Effects. The emphasis here is on assessment of axial dispersion effects for broad molecular weight distribution samples. Following assessment, resolution can be experimentally or computationally corrected, if necessary.

The effects of axial dispersion on both measured concentration (by the DRI) and measured intrinsic viscosity (by the DV and DRI combined) must be examined. To assess the effect on measured concentration, the DRI chromatogram and the "true" calibration curve (determined in the usual way from narrow standards) are used to calculate the molecular weight averages of narrow standards. These averages can then be compared to the true values. Theoretically derived equations for the correction factors show that, within the assumptions involved, the corrections to the averages are independent of the shape or breadth of the chromatogram (2,7-9).

The method of assessing the effect on measured intrinsic viscosity from the DV depends upon an equation derived by Hamielec (10). In the following version of the correction equation, Gaussian spreading of the individual molecular sizes in the chromatograph is assumed.

$$\frac{\overline{[\eta]}(v)}{\overline{[\eta]}(v,uc)} = \frac{F(v)}{F(v - D_{2\eta}(v)\sigma^2(v))} \exp\{\frac{-(D_{2\eta}(v)\sigma)^2}{2}\} \tag{4}$$

where $\overline{[\eta]}(v,uc)$ refers to the value of local intrinsic viscosity obtained from the DV and DRI detectors via Equation (2) and uncorrected for axial dispersion, $D_{2\eta}(v)$ is the slope of the intrinsic viscosity calibration curve (plotted using natural logarithms instead of logarithm to the base ten) and σ is the standard deviation of the Gaussian shape of the chromatogram of a truly monodisperse standard. The right hand side of this equation is a correction factor which can be applied to the measured intrinsic viscosity to convert it to a value corrected for the effects of axial dispersion. It is important to note that this correction factor strongly depends upon the ratio of two different heights of the DRI chromatogram of the sample and therefore upon the shape and breadth of that chromatogram. However, some inaccuracy will be introduced by applying the equation to samples whose composition varies with retention volume because an implicit assumption in development of the equation is that the baseline corrected DRI response is proportional to concentration. The degree of error depends upon how much the detector response actually reflects a composition change with retention volume rather than a total polymer concentration change.

Experimental

Polystyrene standards included NBS 706 (National Bureau of Standards), and PSBR 300K (American Polymer Standards Corp., Ohio). The second standard provided us with a required whole polymer intrinsic viscosity value. Plastic waste used in this work was a clean industrial blend of polyethylene with polypropylene originating from the manufacture of disposable diapers and normally destined for landfill because of its poor mechanical properties. Processing involved adding a free radical initiator (t-butyl cumyl peroxide) and a cross-linking agent (pentaerythritol triacrylate) to the blend during extrusion.

A Waters 150C high temperature SEC operating at 145°C and utilizing 1,2,4-trichlorobenzene as the mobile phase was used. At the exit of the chromatograph, the flow was split between the differential refractometer and a Model 110 differential viscometer (Viscotek Corporation). Four twenty micron Plgel mixed-bed columns were used. Injection concentrations were generally 1.4 to 1.5 mg/mL for polystyrene standards and 1.5 to 2.9 mg/mL for broad molecular weight distributions polyolefins. 0.2 wt. % Irganox 1010 was added as a stabilizer and flow marker for all analyses except polystyrene standards less than 10,000 in peak molecular weight. The differential pressure transducer calibration factor (*1*) ("DPT sensitivity factor") for the differential viscometer was determined as 10/range knob setting when taken at the ten mv output terminals of the DV. Likely a better method, when more standards of known intrinsic viscosity are available, is to use the differential pressure transducer calibration factor which provides the accepted intrinsic viscosity value for a standard (*1,11*).

Results and Discussion

Step 1: Analysis of NBS 706 Using the DRI. Table I shows the results of calculating the molecular weight averages from the DRI chromatogram for NBS 706. "True" values shown for NBS 706 are the most recent estimates obtained from Mourey's work at Eastman Kodak. The deviations observed were tentatively attributed to axial dispersion effects. Considering that the results are from high temperature SEC, they were judged sufficiently good to proceed (at least tentatively) to Step 2.

Table I
Results of Step 1
Molecular Weight Averages of NBS 706

	Calculated	True	% Deviation
$\bar{M}_n$	109000	123000	-11.5
$\bar{M}_w$	277000	276000	0.508
$\bar{M}_z$	516000	435000	18.7

Step 2: Whole Polymer Intrinsic Viscosity of PS. The whole polymer intrinsic viscosity for the polymer standard using the DV and Equation (1) was determined to be 0.77 dl/g. This was in acceptably close agreement with the vendor value of 0.79 dl/g and allowed us to proceed to Step 3.

Step 3: Determination of the Interdetector Volume. Figure 1 shows the superimposition of the local intrinsic viscosity values obtained from measurement by the DV of NBS 706 along with the use of an interdetector volume of 0.104 cc in Equation (2) on the whole polymer values of intrinsic viscosity determined using the DV on each polystyrene standard (Equation (1)). It can be seen that an excellent superposition of the two curves resulted.

Table II shows the molecular weight averages obtained from the molecular weight distribution determined using the DV along with the DRI (Equation (3)). Some improvement over the values shown in Table I were noted. This was interpreted to mean that the "effective" interdetector volume determined was also accomplishing a small amount of resolution correction.

Analysis of Recycled Plastic Waste. Figure 2 shows the normalized chromatogram of the diaper plastic before and after processing. Figure 3 shows a plot of the output of the DV plotted as specific viscosity per unit mass injected versus retention volume. As mentioned above, the area under these plots is the whole polymer intrinsic viscosity. Figure 4 shows the corresponding plots of the local intrinsic viscosities versus retention volume. Using each of these local intrinsic viscosities in Equation (3), the local M_n values were calculated as a function of retention volume and are shown in Figure 5. The most obvious interpretation of these results is that branching occurred across the chromatogram

Table II. Molecular Weight Averages of NBS 706
Using Universal Calibration
(requires: interdetector volume, DV, DRI, $[\overline{\eta}]$ of standards)

	Calculated	True	% Deviation
$\overline{M}_n$	117000	123000	-5.03
			(-11.5)
$\overline{M}_w$	275000	276000	-0.218
			(0.508)
$\overline{M}_z$	504000	435000	15.9
			(18.7)

Note: % Deviations for Step 1 values shown in brackets.

as a result of processing. Measured intrinsic viscosities were consistently less than the starting polymer and molecular weights consequently higher at the same retention volume. However, an assumption in obtaining these results is that the baseline corrected differential refractometer response is proportional to concentration. That is, that the detector response is unaffected by the complexity of the polymer. Factors justifying this assumption are: the similarity of

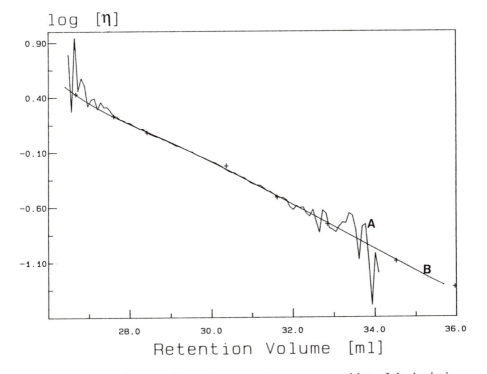

Figure 1: Determination of inter-detector volume: superposition of the intrinsic viscosity calibration curve obtained from DV measurement of NBS 706 at each retention volume (Curve A) superimposed upon that obtained by plotting the whole polymer intrinsic viscosity of narrow standards (Curve B). Peak retention time data for standards are indicated by "+".

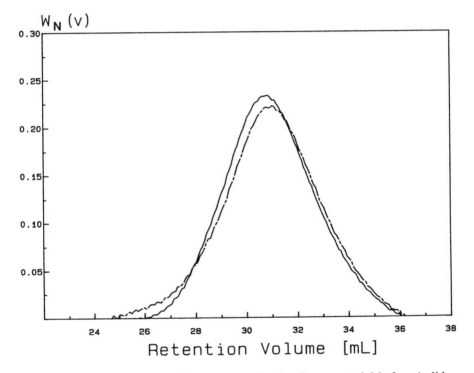

Figure 2: Normalized DRI chromatogram for the diaper material before (solid line) and after (dashed line) processing.

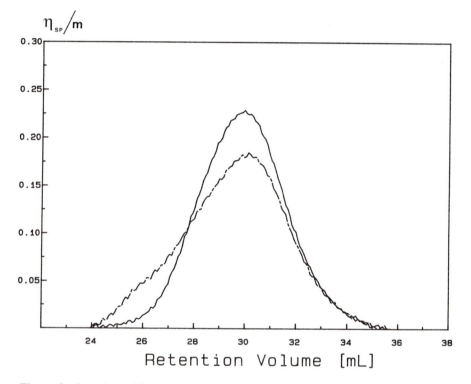

Figure 3: Local specific viscosities divided by mass injected plotted versus retention volume for the diaper material before (solid line) and after (dashed line) processing.

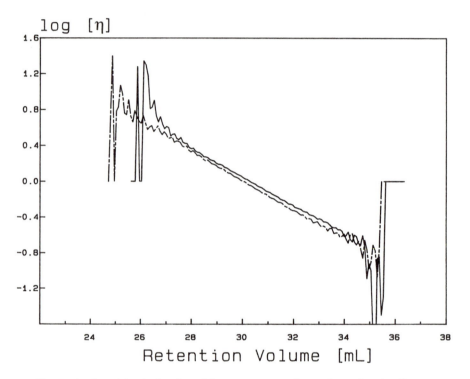

Figure 4: Local intrinsic viscosities versus retention volume for the diaper material before (solid line) and after (dashed line) processing.

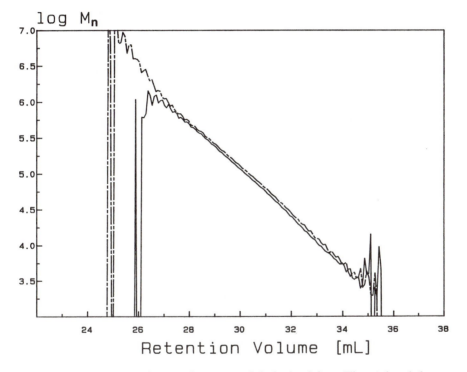

Figure 5: Local M_n for the diaper material obtained from Figure 4 and the universal calibration curve.

polyethylene and polypropylene; the low concentration of cross-linker present in the product; emphasis on seeing differences between the before and after samples. In addition, three main other complications were encountered in this investigation which cause some uncertainty in the results: axial dispersion, low-molecular-weight end resolution and column degradation. These are examined in turn below.

Axial Dispersion. Figure 6 shows a plot of the whole polymer $\bar{M}_n$ and $\bar{M}_w$ values obtained from the chromatograms of narrow standards compared to the values provided by the vendors. Agreement is excellent for $\bar{M}_w$ but much less so for $\bar{M}_n$, particularly at high molecular weights. This asymmetry could be attributed to skewing or to inadequately known $\bar{M}_n$ values (2). Previous experience under similar conditions showed that utilization of the heights of broad DRI chromatograms (i.e. the concentration of individual molecular weights) remained accurate and could be used for process analysis (7-9) while the molecular weight averages could show significant inaccuracy. The main reason for this is that the averages are sensitive to small changes in the chromatogram heights, particularly at the tails.

Figure 7 shows the correction factor required for values of local intrinsic viscosity obtained from the DV as a function of retention volume. NBS706 polystyrene is the sample analyzed. The correction factor is most significant at the tails of the chromatogram and for higher values of standard deviation of the Gaussian spreading function. From measurements on the narrow standards assuming them to be monodisperse, this standard deviation was estimated to be 0.5 mL. Using this pessimistic estimate of the breadth, this means that the correction factor to the local intrinsic viscosities would be approximately 10% at worst. Figure 8 shows the correction factor for a standard deviation of 0.5 mL. calculated from a chromatogram for the diaper plastic waste after processing. The correction is similar to that obtained for the polystyrene sample and may be higher at the extreme low retention volume end. However, noise in the low retention volume region attributed to ratioing the low heights of the chromatogram tail for the calculation is quite severe.

Low Molecular Weight End Resolution. Interference with the low molecular weight end of the chromatogram from the stabilizer peak was a troublesome problem which effectively prohibited calculation of absolute $\bar{M}_n$ values from the method of Goldwasser (12). Various computational attempts to define the tail of the polymer chromatogram based on curve fitting of the stabilizer peak and subtraction from the main peak were not satisfactory. Uncertainty remained high in the low molecular weight tail and a considerable amount of numerical work was required. In more recent work, the problem has been reduced by adding a 50 angstrom column to improve low molecular weight resolution.

Column Degradation. Column degradation affects both the peak retention volumes of standards and the shape of their chromatograms. It is necessary to perform a flow rate correction using the stabilizer as an internal standard on both before arriving at conclusions with regards to column degradation. Figure 9 shows

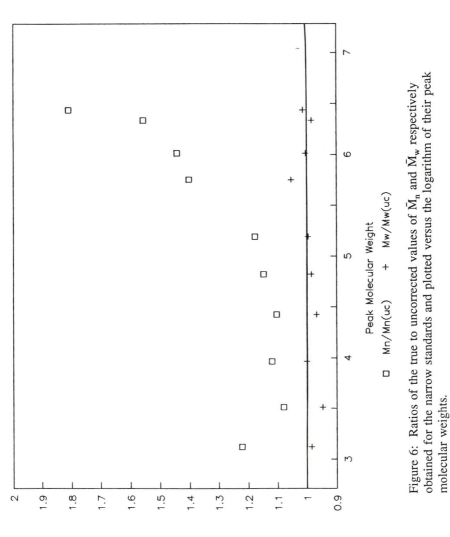

Figure 6: Ratios of the true to uncorrected values of $\bar{M}_n$ and $\bar{M}_w$ respectively obtained for the narrow standards and plotted versus the logarithm of their peak molecular weights.

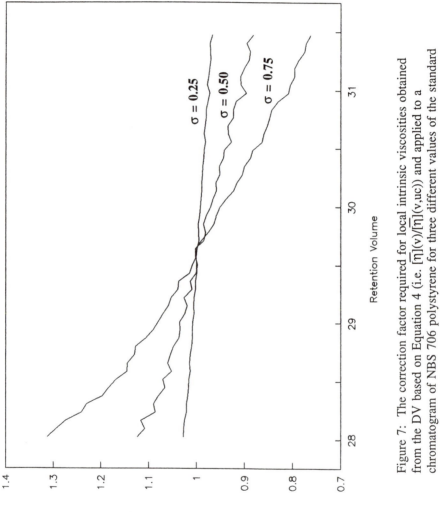

Figure 7: The correction factor required for local intrinsic viscosities obtained from the DV based on Equation 4 (i.e. $[\overline{\eta}](v)/[\overline{\eta}](v,uc)$) and applied to a chromatogram of NBS 706 polystyrene for three different values of the standard deviation of a truly monodisperse polymer.

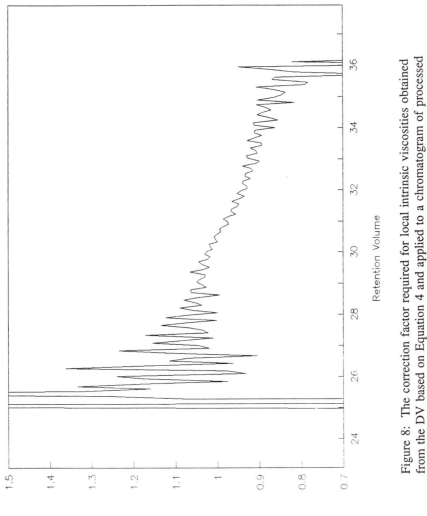

Figure 8: The correction factor required for local intrinsic viscosities obtained from the DV based on Equation 4 and applied to a chromatogram of processed diaper material for a standard deviation of 0.5 mL.

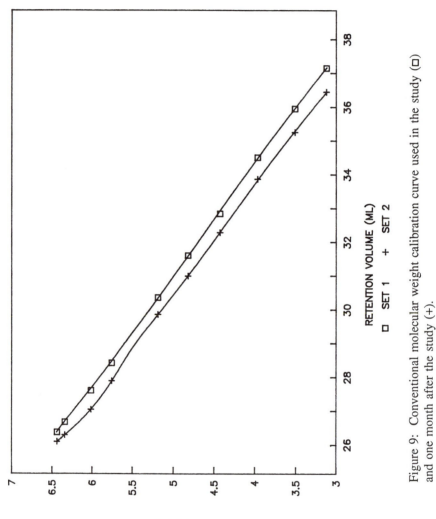

Figure 9: Conventional molecular weight calibration curve used in the study (□) and one month after the study (+).

the calibration curve at the beginning of the study as well as the one about a month after the study was concluded. The lines show the result of a spline fit to the data points. Figure 10 shows the latter curve after flow rate correction utilizing the internal standard. It is evident that, with the exception of the very low retention volume end, the curves of Figure 9 superimposed by the correction. However, there is a possibility that the apparent flow rate change was actually an effect of column degradation. Figures 11 and 12 show chromatograms of the standards corresponding to the early and late calibration curves respectively. Figure 13 shows these chromatograms after flow rate correction. The curves obtained later show obvious broadening as a result of column degradation. This would have remained concealed if only flow-rate-corrected peak retention volumes were examined.

Conclusions

The utility of the systematic approach for multidetector SEC interpretation was demonstrated in a very non-ideal situation involving analysis of recycled waste plastic in a high temperature SEC (*1*).

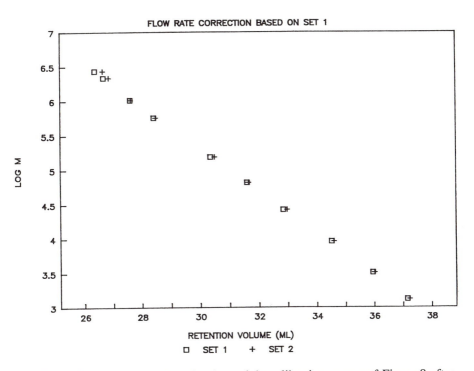

Figure 10: Conventional molecular weight calibration curves of Figure 9 after flow rate correction.

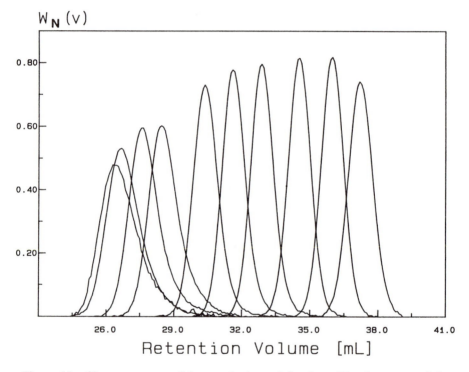

Figure 11: Chromatograms of the standards used for the calibration curve of the study.

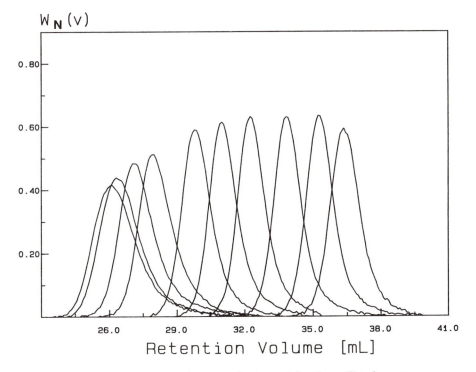

Figure 12: Chromatograms of the standards used for the calibration curve one month after the study.

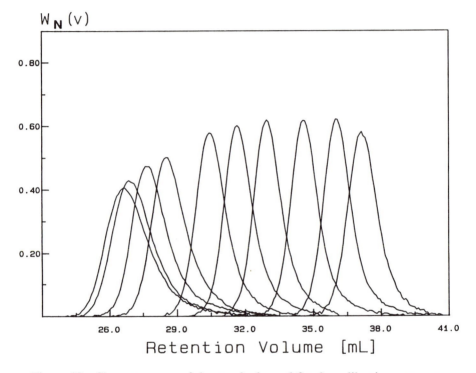

Figure 13: Chromatograms of the standards used for the calibration curve one month after the study and flow rate corrected.

Axial dispersion assessment should be considered as a necessary part of the systematic approach.

Column degradation should be examined using flow-rate-corrected chromatograms of narrow standards. If a change in shape is observed then the apparent flow rate change can probably be attributed to column degradation.

Low molecular weight interference was found to be best dealt with experimentally.

Acknowledgments

We are grateful to the Eastman Kodak Company, the Ontario Centre for Materials Research and the Natural Sciences and Engineering Research Council of Canada for their financial support of this project. Also, we wish to thank Resource Plastics Corporation, Himont Canada, Atochem North America and Viscotek Corporation for their assistance.

Literature Cited

(1) Mourey, T. H.; Balke; S. T. "A Strategy for Interpreting Multi-Detector SEC Data I. Development of a Systematic Approach", this volume
(2) Balke, S.T. *Quantitative Column Liquid Chromatography, A Survey of Chemometric Methods;* J. of Chromatogr. Library; Elsevier: NY, NY, **1984**; Vol. 29.
(3) Hamielec, A. E.; Ouano, A.C. *J. Liq. Chromatogr.*, **1978**, *1*, pp. 111-120.
(4) Balke, S. T.; Cheung, P.; Lew, R.; Mourey, T. H.; *J. Liq. Chromatogr.*, **1990**, *13*, pp. 2929-2955.
(5) Cheung, P.; Balke, S. T.; Mourey, T. H. *J. Liq. Chromatogr.*, **1992**, *15*, pp. 39-69.
(6) Mourey, T. H.; Miller, S. M. *J. Liq. Chromatogr.*, **1990**, *13*, pp. 693-702.
(7) Balke S.T.; Patel, R.D.; in *Size Exclusion Chromatography*; Provder, T., Ed.; ACS Symp. Ser., **1980**, Vol. 138, pp. 149-182.
(8) Lew, R.; Suwanda, D.; Balke, S. T. *J. Appl. Polym. Sci.*, **1988**, *35*, pp. 1049-1063.
(9) Lew, R.; Cheung, P.; Suwanda, D.; Balke, S.T. *J. Appl. Polym. Sci.*, **1988**, *35*, pp. 1065-1084.
(10) Hamielec, A. E. *Pure and Appl. Chem.*, **1982**, *54*, pp. 293-307.
(11) Styring, M. G., Armonas, J. E., Hamielec, A. E., *J. Liquid Chromatogr.*, **1987**, *10*, pp. 783-804.
(12) Goldwasser, J.M. in *Proc. Int. Gel Permeation Chromatography Symposium*, Newton, MA, **1989**, pp. 150-157.

RECEIVED October 6, 1992

Chapter 14

Single-Capillary Viscometer Used for Accurate Determination of Molecular Weights and Mark—Houwink Constants

James Lesec[1], Michele Millequant[1], and Trevor Havard[2]

[1]**Centre National de la Recherche Scientifique, ESPCI, 10 rue Vauquelin, 75231, Paris 05, France**
[2]**Millipore-Waters, 34 Maple Street, Milford, MA 01757**

The Waters 150 GPC/Viscometry system is equipped with an on-line single capillary viscometer. The geometrical characteristics of the viscometer (18 µl internal volume, 2800 Sec^{-1} shear rate) allow measurements to be made near classical Ubbelohde viscometry conditions. When used in conjunction with a GPC/Viscometry software, it is possible to use universal calibration and to measure real molecular weights. K and Mark-Houwink constants can also be determined and, consequently, long-chain branching g' distribution can be studied for branched polymers. The first trials have shown that a weak flow fluctuation was occurring when the samples pass through the detector due to an increase of specific viscosity by detector constriction. The consequence of this flow fluctuation is a fluctuation of viscometer baseline leading to a small peak deformation looking exactly like a downstream peak shift. At the same time, the viscosity law rotates a little giving a small decrease of value. To avoid this drawback, a new refractometer prototype was built to accommodate this design change.

The development of on-line viscometry detectors for gel permeation chromatography (GPC) has focused attention towards the use of information derived from a GPC/Viscometry system to calculate accurate (absolute) molecular weight averages and branching information. The objectives of this paper are to examine how accurate the slice intrinsic viscosity of a polymer distribution is for determining K and alpha and how design aspects of the system contribute to the accuracy of data obtained. This technique will be applied to commonly available narrow and broad distribution polymers. The results were obtained using a single capillary viscometry detector in series with a modified design of the high sensitivity differential refractometer integrated into the 150CV GPC system. The incorporation of a viscometry detector into a GPC system enables intrinsic viscosity slices to be calculated for the polymer molecular weight distribution. The use of a single capillary viscometer for calculating the intrinsic viscosity information has been documented by Ouano *(1)*, Lesec *(2)* & Provder *(3)*. The ability of the 150CV system has been demonstrated to provide accurate molecular weight averages and bulk intrinsic viscosity values for known polymers. Using this system, excellent results for K and alpha values can be obtained by generating the universal calibration for a range of narrow standards using a Mark-Houwink (Viscosity Law) plot, Figure 1. The calculation of these values is derived from using the supplied peak molecular weight data for the given standards and then calculating the bulk intrinsic viscosity of each standard directly from the viscosity detector. The ability to calculate the slice intrinsic viscosity for an unknown polymer is dependent upon a number of assumptions and extra parameters such as the determination of detector offset.

Parameters That Affect the Calculation of Slice Intrinsic Viscosity

The calculation of slice intrinsic viscosity is a function of both the refractometer and the viscometry detector and careful control of the GPC system parameters. The relationship of the data obtained from the viscometry detector to the slice intrinsic viscosity can be expressed by the following equation:

$$\eta^*_i = [\eta]_i \, C_i \tag{1}$$

$[\eta]_i$ is the slice intrinsic viscosity. C_i is the concentration of the viscosity slice. η^*_i is derived directly from the viscometry detector using the information from a single differential reluctance transducer.

$$\eta^*_i = [2\,((P_i - P_o)/\,P_o - \ln P_i\,/P_o)\,]^{0.5} \tag{2}$$

P_i is the pressure from the sample and solvent at a constant flow in the capillary. P_o is the pressure of solvent in the capillary.

C_i is derived directly from the GPC system.

$$C_i = (A_{N\,i} \; Conc. \; Inj \; Vol.)\,/\,\Delta\,V \tag{3}$$

GPC system sample concentration

$$C = \Sigma\,C_i \tag{4}$$

A_{Ni} is the area normalized slice from the differential refractometer distribution. Conc. is the concentration of the injected sample.

Inj Vol. is the true injected volume. ΔV is the absolute volume increment of the intrinsic viscosity slice.

Once the concentration of the viscosity slice has been determined, the slice intrinsic viscosity can then be calculated. This calculation assumes that the offset between the two detectors is known. The sample intrinsic viscosity can now be calculated using equation 5.

$$[\eta] = \Sigma\; [\eta]_i\,C_i\,/\,\Sigma\,C_i \tag{5}$$

In order to calculate the sample intrinsic viscosity without the use of the differential refractometer, we can substitute equations 1 and 4 into equation 5 to obtain equation 6.

$$[\eta] = \Sigma \, \eta^*_i \, / \, C \tag{6}$$

The use of these equations to calculate the intrinsic viscosity by different routes gives diagnostic information concerning the GPC/Viscometry system. Equation 5 demonstrates that the two detectors and the GPC system provide the correct sample intrinsic viscosity independent of any correction factors. With knowledge of the injection concentration and viscometry detector information, the correct sample intrinsic viscosity can be determined independently of the differential refractometer signal, equation 6. Using the above calculations to evaluate the viscometer performance can be misleading as GPC/Viscometry is not only used for the determination of sample intrinsic viscosity and accurate molecular weight averages, but also to determine the K and alpha (a) values from the Mark-Houwink relationship.   The universal calibration provides an excellent demonstration of how the Mark-Houwink relationship, Figure 1, can be derived from using the narrow standard peak molecular weight values supplied from the manufacturer and the calculated standards intrinsic viscosity from the GPC/Viscometry system.

$$Log\,[\eta] = Log\,K + aLog\,M \tag{7}$$

The universal calibration, Figure 2, provides the means by which accurate molecular weights can be determined for unknown samples.

The Mark-Houwink relationship, Figure 1, enables the user to determine how accurate the calibration is in terms of the supplied peak molecular weight and calculated intrinsic viscosity for the narrow standards. The advantage of the K and alpha values determined from the calibration curve is that the values are discreet for each narrow standard and the intrinsic viscosity of the standards can be calculated directly from equation 6. When comparing the Mark-Houwink plot of broad standard polystyrene standards to the values obtained from the universal calibration, we assume that we know the correct detector offset between the differential and viscometry detectors. Provder *(3)* found that the measured values for detector offset do not match the expected value from a measured offset using a low molecular marker peak like toluene.   (Table I)

Table I. Effect of Dead Volume Between Detectors
(Test Sample: Dow 1683 Polystyrene)

Vol (μl)		M_n x10^{-3}	M_w x10^{-3}	$[\eta]$ (dl/g)	K x10^{-4}	α
1.64	(0.1 sec)	100.2	250.2	0.856	2.21	0.670
19.0	(1.2 sec)	100.5	249.7	0.856	1.65	0.693
24.6	(1.5 sec)	101.6	249.8	0.856	1.47	0.702
49.2	(3.0 sec)	101.6	249.1	0.854	1.05	0.730
79.0	(4.8 sec)	106.7	248.8	0.854	0.71	0.761

For the correct K and alpha values for the Dow polystyrene sample, a 1.5

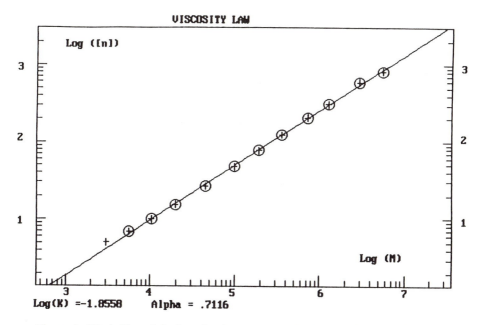

Figure 1. Mark-Houwink plot of polystyrene standards from 2,000 to 5 million molecular weight using 3 µstyragel HT columns.

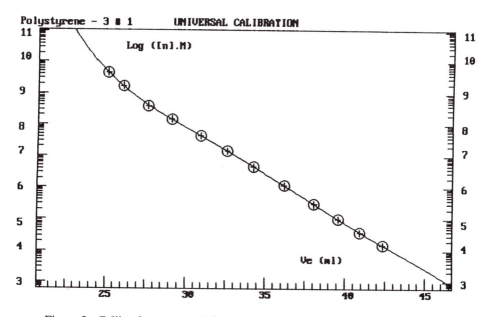

Figure 2. Calibration generated for the polystyrene standards using 2,000 to 5 million molecular weight narrow standards.

second offset was used. The measured offset using a marker peak was 4.8 seconds. Once the offset is corrected for the broad sample, another assumption is made that the corrected value is true for all other samples. In order to determine whether or not this is true, the authors carried out a number of experiments to determine if the difference between the measured offset and the corrected peak shift is a function of peak shift or some other phenomena. The viscometry detector is expected to respond to the eluting high molecular weight sample before the differential refractometer, and the differential refractometer should be more sensitive to the low molecular weight portion of the distribution than the viscometry detector. In some cases, the response from the viscometry detector appeared to have low molecular weight tail for the samples even when there was little or no response from the differential refractometer. A possible explanation for this effect is that there are minor flow rate variations as the sample elutes from the columns through the detectors. An experiment was carried out to determine if this explanation had any validity.

Experimental

200 µl of a TSK 2 million molecular weight polystyrene narrow standard (0.1%) was injected into a production model Waters 150CV system using 2 ultrastyragel columns; THF as the eluent; and a flow rate of 1mL/min. Several feet of 0.040 inch (I.D.) tubing was placed between the viscometer and the differential refractometer. Large diameter tubing was used only in order to remove any constriction between the two detectors during elution through the viscometer capillary. This experiment is carried out to evaluate flow effects due to constriction. The tubing was removed for all GPC experiments.

Results

A comparison of the raw data obtained for the 2 million molecular weight sample with and without downstream constriction demonstrates an obvious peak shift in the viscometer capillary. (Figure 3)

By subtracting the raw data file without constriction from the data obtained on the production model, it is possible to visualize this flow effect. This slow down and speed up of the flow provides significant error for every experimental viscosity data slice except where the flow curve meets the apparent baseline. Therefore, an incorrect value for the slice intrinsic viscosity will be calculated. This effect can occur because there is constriction in between or in the detector itself and may also be affected by the type of constriction used.

The flow variation created by the constriction of 0.009 inch (I.D.) tubing under normal operating conditions is very small, less than 0.02% RSD, which is less than the specified precision for most standard HPLC pumps used in a GPC system. (Figure 4)

Although this effect will produce significant error for the determination of slice intrinsic viscosity and the Mark-Houwink K and alpha values, the calculation of the molecular weight averages appears to be unaffected. **Therefore, whether there is constriction in the GPC system or not, the viscometry detector will provide accurate molecular weights**. Exceptions to the rule may be very high molecular weight polymers where the specific viscosity without decreasing the concentration of the material will amplify flow effect to such a level that this will lead to significant error.

Viscometer With & Without Downstream Constriction

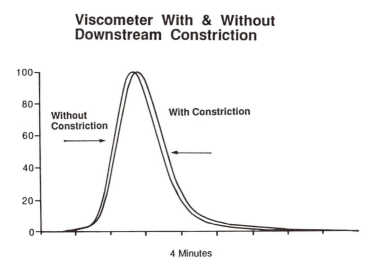

Figure 3. Amplification of the flow effect using 0.2% solution of a 2 million molecular weight narrow polystyrene standard on 2 linear ultrastyragel columns.

Comparison of Amplified Lesec Effect to a Normal Distribution

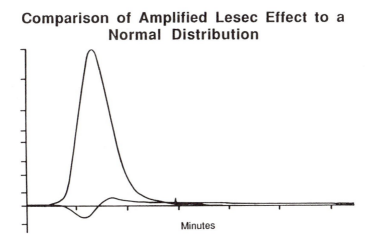

Figure 4. By subtracting the chromatogram in Figure 3 it is possible to demonstrate the flow effect. This effect is caused by excess constriction from 0.09 inch ID tubing.

Solution to the Problem of Flow Effects Between the Columns and the Detectors

Waters has built a prototype differential refractometer in an attempt to eliminate or significantly reduce the constriction between the viscometry detector and the differential refractometer.

For the polystyrene narrow standard, the experimental log intrinsic viscosity data versus elution volume for the current 150CV unit is plotted, Figure 5. All of the experimental data points were fitted to a 3rd order fit and used to calculate the molecular weight and Mark-Houwink relationship. The 10300 Daltons molecular weight standard, when treated as an unknown, gave peak molecular weight of 10590 Daltons for standard GPC and 10630 Daltons for GPC/Viscometry; demonstrating that the molecular weight calculation is unaffected by the flow variation during elution through the detectors. The detector offset used was 80 µl which is the measured geometrical offset between the two detectors. The K and alpha values are incorrect. Where the alpha value in this case is a negative value, this means that graphically the plotted molecular weight distribution will be incorrect. The current differential refractometer design was replaced by a new design which eliminates the constriction. The correct geometrical offset of 145 µl was used for the detectors (4). The change in experiment log intrinsic viscosity is now plotted versus elution volume with the correct slope and a more linear curve for the experimental data points for the same sample. (Figure 6)

The calculated molecular weight averages and the Mark-Houwink constants are correctly calculated. An interesting observation is that the 3rd order experimental data extrapolation was used over a very wide range in the distribution, demonstrating the high sensitivity of both the viscometry detector and the differential refractometer. The new refractometer design enables log K and alpha values to be calculated. These values approach the values obtained using the Mark-Houwink plot from the universal calibration. In addition, a comparison is shown to demonstrate the effect of axial dispersion on the calculation of these values. (Figure 7, Table II)

The plot range for each narrow standard is very close to the plot obtained from the universal calibration. All of the standards analyzed were obtained from TSK.

The new refractometer design, when used with broad polymer samples, produced excellent results. The log K and alpha values, when averaged for five different samples, gave extremely close agreement to the values obtained from the calibration curve. (Figure 8, Table III)

The overlay of the Mark-Houwink plots again demonstrates that these values can be correctly obtained by using the geometrical offset. It is important to recognize the importance of the confidence of these results; otherwise, the use of viscometry detectors for the prediction of branching data and radius of gyration calculations will be misleading, as these calculations are dependent upon the ability to calculate the correct slice intrinsic viscosity distribution for any unknown polymer.

Conclusion

There are flow variations that occur in GPC systems. The use of excessive restriction between columns and detectors will cause flow variations. These variations do not appear to effect the calculation of the molecular weight averages.

**Experimental Data for 10.3 K
Using the Current 150 CV**

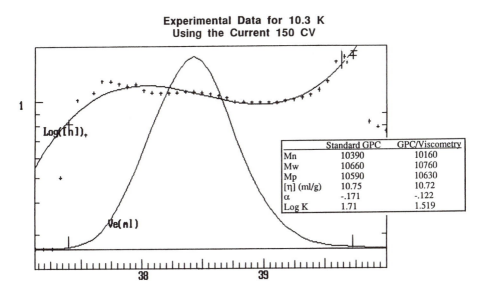

	Standard GPC	GPC/Viscometry
Mn	10390	10160
Mw	10660	10760
Mp	10590	10630
[η] (ml/g)	10.75	10.72
α	-.171	-.122
Log K	1.71	1.519

Figure 5. The experimental log [η] is distorted by the flow effect but does not significantly affect the molecular weight averages.

**Experimental Data for 10.3 K
Using the 150CV with the Prototype
Refractometer**

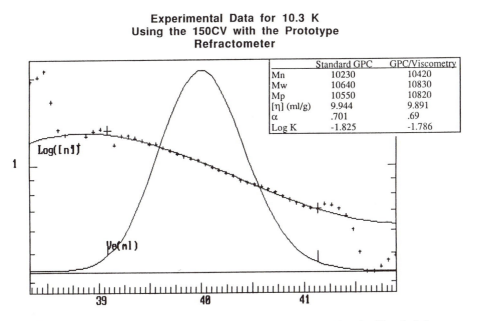

	Standard GPC	GPC/Viscometry
Mn	10230	10420
Mw	10640	10830
Mp	10550	10820
[η] (ml/g)	9.944	9.891
α	.701	.69
Log K	-1.825	-1.786

Figure 6. Using the new refractometer system design enables the K and alpha values to be accurately calculated as well as the molecular weights.

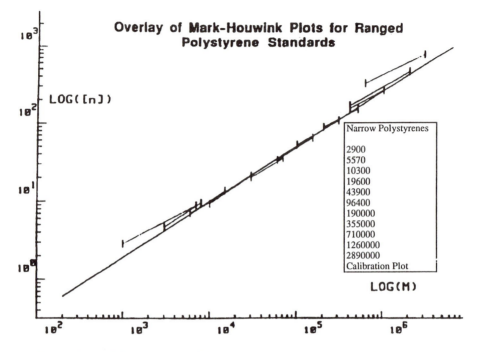

Figure 7. The Mark-Houwink plot demonstrates the accuracy of the new design to provide useful intrinsic viscosity distributions for polymers of narrow polydispersity.

Table II. Polystyrene Narrow Standards Using Prototype 150CV Refractometer Design

Axial Dispersion Molecular Weight	(Uncorrected)		(Corrected)	
	Alpha	Log K	Alpha	Log K
2900	0.55	-1.18	0.58	-1.20
5570	0.69	-1.75	0.73	-1.88
10300	0.73	-1.95	0.72	-1.87
19600	0.72	-1.95	0.74	-1.99
43000	0.60	-1.34	0.66	-1.60
96000	0.70	-1.78	0.70	-1.85
190000	0.53	-0.89	0.66	-1.60
335000	0.49	-0.59	0.59	-1.17
710000	0.43	-0.21	0.59	-1.15
1260000	0.55	-0.71	0.63	-1.33
2890000	0.54	-0.67	0.54	-0.64

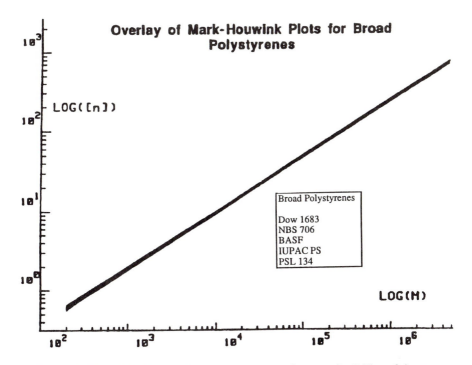

Figure 8. The Mark-Houwink plot demonstrates the reproducibility of the new refractometer system design for a variety of broad polystyrenes.

Table III. Polystyrene Broad Polymers Using Prototype 150CV Refractometer Design

Polymer Type	Mw	Mn	Alpha	Log K
Dow 1683	248,000	99,810	0.712	-1.878
NBS 706	254,000	94,500	0.708	-1.836
BASF 168N	316,900	138,000	0.716	-1.899
PS IUPAC	227,800	54,000	0.719	-1.910
PSL 134	231,000	97,650	0.705	-1.856
Polymer Type Average			0.712	-1.876
Calibration			0.712	-1.856

By moving the detector offset to compensate for the flow variation, the calculation of the Mark-Houwink values can be calculated, yet the offset may not remain constant for all unknown polymers. In the Waters 150CV system where all of the components essential to GPC/Viscometry are plumbed in series, the speed up and slow down of flow, if tested using a marker peak for long term flow precision, cannot be detected. The use of the new design of differential refractometer enabled the accurate determination of slice intrinsic viscosity data for broad polymer samples as well as providing a system that is able to analyze narrow distribution polymer without the need for any correction factors unrelated to measured physical parameters. The prototype provided improved data for slice intrinsic viscosity as the flow effect had been eliminated. This flow effect not only applies to the 150CV but will also apply to any system that incorporates any type of constriction to balance flow splitting, reservoirs, connecting columns to detectors, or detectors to detectors.

Literature Cited

1. Ouano, A. C. *J. O. Polymer Sci.* **1972**, *Vol 10*, Part A-1, pp 2167-2180.
2. Lesec, J.; Lechacheux, D.; Marot, G.; *International GPC Symposium '87 Proceedings,* Waters, Division of Millipore, Milford, MA, 1987; pp 81-112.
3. Kuo, C.; Provder, T.; Koehler, M. E.; *International GPC Symposium '89 Proceedings,* Waters, Division of Millipore, Milford, MA, 1989; pp 68.
4. Lesec, J.; Havard, T.; Method of offset, future publication.

RECEIVED October 22, 1992

Chapter 15

Gel Permeation Chromatography—Viscometry
Various Solvent Systems

C. Kuo, Theodore Provder, and M. E. Koehler

Research Center, The Glidden Company, 16651 Sprague Road,
Strongsville, OH 44136–1739

The characterization of molecular weight distribution, intrinsic viscosity and polymer chain branching as a function of molecular weight for polymer standards and polymers of commercial interest are reported for tetrahydrofuran (THF), dimethylformamide (DMF) and dimethylacetamide (DMAC) mobile phases. The instrument used in this study is a single capillary GPC - Viscometer/Data Analysis System. In this chapter additional studies on operational variables, particularly dead volume, is reported. The validity of universal calibration in DMF and DMAC is discussed for a range of polymer types, and column packings.

Recent developments in gel permeation chromatography (GPC) have been focused on molecular size sensitive detectors in the form of light scattering detectors and viscometer detectors for the determination of absolute molecular weight distribution and polymer chain branching. Commercially available GPC viscometers were introduced by Viscotek in 1984 (1) and by Millipore Waters Chromatography Division in 1989 (2). In a previous publication (3) the principle of operation, instrumentation, operational variable considerations and data analysis methodology and its application to polymer systems in THF were described and discussed for a commercial single capillary GPC-Viscometer System. In this chapter the effect of dead volume on viscosity law parameters is further elaborated and comparative molecular weight results in THF are presented for linear and randomly branched polystyrene polymers.

The practice of GPC in highly polar solvents such as DMF and DMAC is not fully understood with respect to polymer-solvent-columns packing interactions. These interactions often prevent validated universal calibration curves from being generated which in turn, subsequently, prevent accurate molecular weight distribution information from being attained from viscometer detector data for a variety of polymer types. Another related effect which hinders accurate quantitation of GPC data in highly polar solvent systems is the presence of solvent trash peaks in the molecular weight range of interest. In this chapter experimental conditions for minimizing the effect of solvent trash peaks are explored. The validity of universal calibration in conjunction with viscometer detection is examined for various column

0097–6156/93/0521–0231$06.00/0

packing types in DMF and DMAC based solvent systems for a variety of polymer types.

Experimental

(A) Instruments: (1) Millipore Waters Single Capillary GPC-Viscometer/Data Analysis System (GPCV)(2); (2) Glidden GPC Viscometer/Data Analysis System (GPC/VIS(3))

(B) Columns: (1) Millipore Waters Ultrastyragel Columns with 10^3, 10^4, 10^5, 10^6 Å porosity; (2) Millipore Waters μstyragel HT Linear columns (10 μm);(3) Polymer Laboratories PLgel columns (5 μm) with 10^5, 10^3, and 50 Å porosity;(4) Millipore Waters Ultrahydrogel linear column (6-13μm); (5) Shodex KD-802.5 (7 μm), (6) Shodex KB-802.5(7μm)(hydroxyl functional packing) (7) Millipore Waters Prototype μHT.

(C) Mobile Phases: (1) THF; (2) DMF; (3) DMF/0.1M LiBr; (4) DMAC.

(D) Calibrants: (1) Narrow molecular weight distribution (MWD) polystyrene standards (Toyo Soda Co.); (2) Narrow MWD polymethyl methacrylate standards (American Polymer Standards Corp.); (3) Polyethylene Oxide/Polypropylene Glycol (PEO/PPG) (American Polymer Standards Corp.).

(E) Materials: (1) Broad MWD PS: Dow 1683, NBS 706, ASTM PS4; (2) Broad MWD PMMA: Polymer Bank 6041, Aldrich 18226-5; (3) PVC: Pressure Chemical PV-4; (4) Polyvinyl acetate (PVAc): Cellomer 024 CO1 and 024 CO3; (5) Branched PS: Branch B (The University of Akron).

Results and Discussion

Ultrastyragel GPCV/THF. Dead Volume (Viscometer Delay Time): As shown in the previous paper (3) the dead volume difference between the viscometer and the DRI detectors must be accounted for. Otherwise, systematic errors in the Mark-Houwink parameters K and α can occur. Table I shows the effect of dead volume on the molecular weight averages, the intrinsic viscosity and the Mark-Houwink parameters. As we reported previously (4,5) K and α are very sensitive to the value of the dead volume between detectors. However, the molecular weight averages and the bulk intrinsic viscosity are barely affected. The viscometer delay time was estimated to be 1.5 seconds (24.6 μl) by matching K and α values to those obtained from the viscosity law plot from narrow molecular weight distribution polystyrene standards by on line GPC-viscometry (K = 1.5 x 10^{-4} and α = 0.702). The measured delay time between detectors using toluene was 4.8 seconds (79 μl) which is close to the value calculatable from the physical dimensions of the tubing. The cause of the discrepancy, known as the "LeSec Effect", and the remedy to correct it was the subject of a recent study (6) reported in the First International GPC/Viscometry Symposium held at Del Lago Resort, TX, April 24-26, 1991. This "Lesec Effect" also was investigated in this laboratory through a modified refractometer where the original connecting tubing was replaced with larger ID tubing resulting in a larger dead volume (300 μl). In the process of calibrating this modified refractometer with narrow MWD polystyrene standards, a set of reasonable K and α values (0.00019 and 0.69) is obtained. However, the results obtained for the broad MWD polystyrenes (NBS 706 and Dow 1683) indicated the α values are highly overestimated. Even with zero dead volume, the α value is still as high as 1.00. Apparently, negative dead volume values had to be used to obtain a reasonable α value. This confirms our earlier observation that the GPCV software is treating the dead volume different from Glidden GPC/VIS (3), and indicates an inadvertent minus sign error in the Millipore-Waters GPCV software.

Linear and Branched Polymers. The results of a series of commercially available polymers analyzed with this system have been reported in the previous paper (3). In this chapter, attention will be focused on a randomly branched polystyrene sample which was obtained from the University of Akron. The results are listed in Table II along with those of the linear NBS 706. It is seen that although the molecular weight of this branched sample is equivalent to that of the linear NBS 706, viscosity is lower. Upon comparing the log [η] vs. log M (viscosity law) plot in Figure 1 the indication of branching is evidenced by the deviation of the viscosity from the linear NBS 706 polystyrene polymer in the higher MW regions. The branching index can be obtained as a function of molecular weight by ratioing the viscosity of the branched polystyrene sample to that of the linear polystyrene sample at the same molecular weight.

GPCV/DMF and GPCV/DMF (0.1M LiBr)

To provide MWD information for polymers which are not THF soluble, DMF was investigated as a GPC eluant. Various combinations of columns were explored. One major problem in using DMF with the crosslinked polystyrene gel packing is the interference of the solvent trash peaks with the low molecular weight (MW) polystyrene standards as shown in Figure 2 for three PLgel columns (10^5, 10^3, 50 Å). It is seen that the polystyrene 500 MW standard eluted after the solvent peak which also interferes with the polystyrene 2100 MW standard. The apparent retardation of elution of low MW polystyrene molecules was explained by earlier workers (7,8) in terms of adsorption of the polystyrene on the apolar polystyrene gel packings. Addition of LiBr modifier did not improve but further retarded the elution time. Similar limitations in the low MW region were observed for a column set consisting of 3 μHT linear columns plus a Shodex KD-802.5 column.

By using a Ultrahydrogel column (PMMA-type gel packing) with the 3 μHT linear columns, the solvent trash peak elutes much later in time as shown in Figure 3 for the polystyrene 1350 standard. With the addition of the Ultrahydrogel column, the low MW region can be extended to resolving the polystyrene 800 MW standard. This will provide for improved quantification of molecular weight distribution statistics in the low molecular weight region. This column set was calibrated with the narrow MWD polystyrene standards with the GPCV data system.

Figure 4a shows the hydrodynamic volume calibration curve and Figure 4b shows the viscosity law plot for the generation of the Mark-Houwink parameters K and α. These values are in good agreement with the literature data (9,10) A series of broad MWD polymers were run on this system. The results are shown in Table III. In general, the molecular weight averages and bulk intrinsic viscosity are in reasonable agreement with the nominal values (11), although the number average molecular weight seems to be higher than those obtained with the Ultrastyragel/THF system. This could be explained by the lack of resolution at the low MW regions. The system also can detect the branching of a polyvinyl acetate sample as shown in Figure 5 by the branching index plot.

We also examined the PLgel column set with DMF/0.1 M LiBr as the eluant. Using narrow MWD PMMA standards as calibrants, all the MW information obtained for the broad MWD samples seem to be underestimated. The exact cause of the problem is unclear at this time, although we did observe higher system pressure. In a recent paper (12) in a similar study, it was mentioned that smaller particle size (5 μm) PLgel columns might not be suitable for viscous eluants such as DMF due to high operation pressure and shear degradation of high MW polymers.

GPCV/DMAC

Since the validity of universal calibration in DMF may be questionable for a wide

TABLE I. EFFECT OF DEAD VOLUME BETWEEN DETECTORS
(TEST SAMPLE: DOW 1683 POLYSTYRENE)

Vol (μl)		$\overline{Mn}$ x10^{-3}	$\overline{Mw}$ x10^{-3}	$[\eta]$ (dl/g)	K x10^4	α
1.64	(0.1 sec)	100.2	250.2	0.856	2.21	0.670
19.0	(1.2 sec)	100.5	249.7	0.856	1.65	0.693
24.6	(1.5 sec)	101.6	249.8	0.856	1.47	0.702
49.2	(3.0 sec)	101.6	249.1	0.854	1.05	0.730
79.0	(4.8 sec)	106.7	248.8	0.854	0.71	0.761

TABLE II. GPC VISCOMETER RESULTS FOR BROAD MWD
LINEAR AND BRANCHED STYRENE SAMPLES

Sample	$\overline{Mn}$ x10^{-3}	$\overline{Mw}$ x10^{-3}	$[\eta]$ (dl/g)	K x10^4	α
NBS 706	100	263	0.93	1.6	0.69
Branched B	132	269	0.84	2.1	0.63

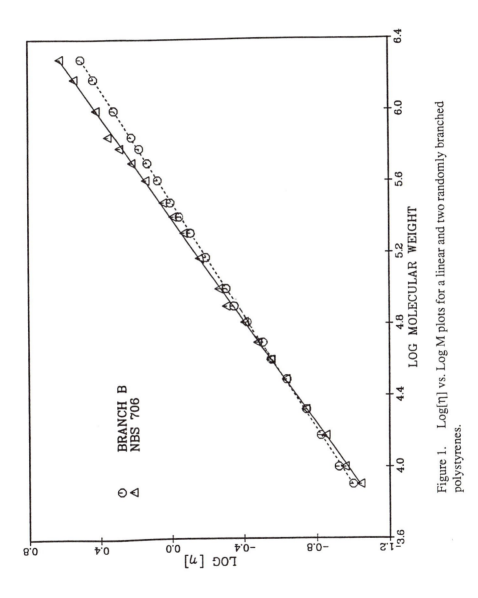

Figure 1.   Log[η] vs. Log M plots for a linear and two randomly branched polystyrenes.

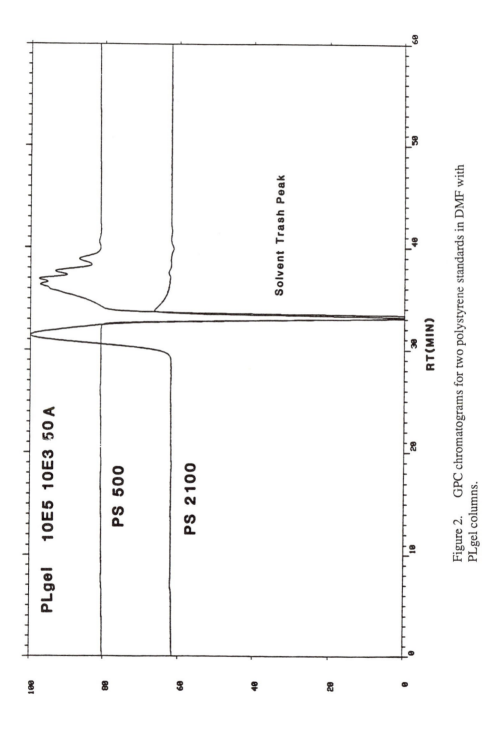

Figure 2.   GPC chromatograms for two polystyrene standards in DMF with PLgel columns.

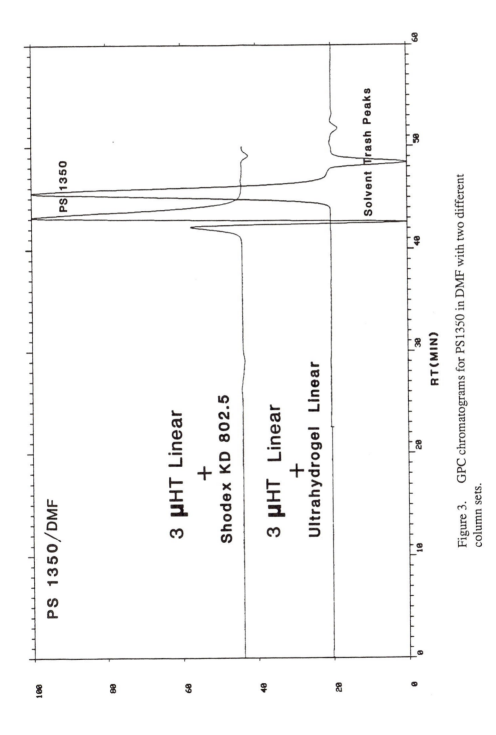

Figure 3.   GPC chromatograms for PS1350 in DMF with two different column sets.

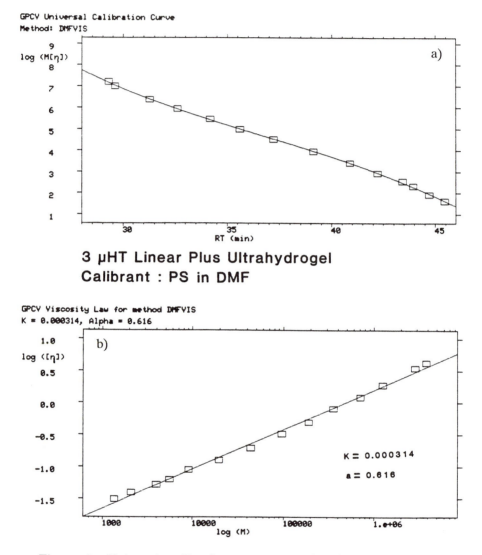

Figure 4. Universal calibration curve and viscosity law plot for polystyrene in DMF.

TABLE III. GPCV/DMF RESULTS FOR BROAD MWD POLYMERS
(3μHT LINEAR & ULTRAHYDROGEL LINEAR/DMF)

Sample	$\overline{Mn}$ x10^{-3}	$\overline{Mw}$ x10^{-3}	$[\eta]$ (dl/g)	K x10^4	α
Dow 1683					
Nominal Values	100	250	-	3.18[9]	0.603[9]
	-	-	-	2.80[10]	0.606[10]
GPCV/DMF	101	225	0.57	2.41	0.637
NBS 706					
Nominal Values	136	258	0.557[11]	-	-
GPCV/DMF	144	260	0.597	0.65	0.74
Eastman 6041 PMMA					
Nominal Values	160	267	-	-	-
GPCV/DMF	175	243	0.65	-	-

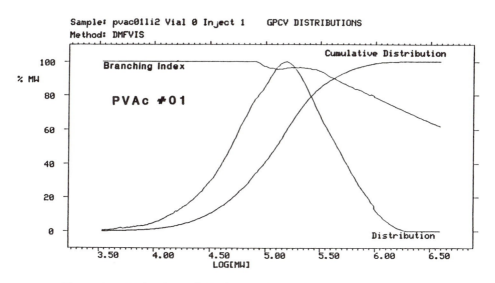

Figure 5.   Molecular weight distribution and branching index plots for
polyvinyl acetate in DMF.

variety of polymers due to the adsorption of polymers on the apolar polystyrene gel, dimethyl acetamide (DMAC) was explored as an alternative solvent. The results indicate that with 3 μHT linear columns and a Shodex KD-802.5 column set, the universal calibration concept was applicable. Figure 6 shows the primary molecular weight calibration curves for PS, PMMA and PEO/PPG standards. As expected, POE/PPG did not fall into the same curve as those of PS and PMMA. However, when the data are plotted onto a universal calibration curve, it all falls onto a common line as shown in Figure 7. The K and α values obtained from the narrow MWD polystyrene standards are 0.000129 and 0.696, respectively. These values compared favorably with the literature data (13). The results obtained for the two broad MWD polystyrenes are listed in Table IV.

Conclusions

The single capillary viscometer hardware functions well and exhibits good baseline stability. For broad MWD samples, accurate absolute molecular weight averages, bulk intrinsic viscosity values, and Mark-Houwink K and α parameters can be obtained from a single GPC experiment. In addition, the software provides branching information as a function of MW for the branched polymers. GPC-viscometry studies in DMF or DMF/LiBr solvent indicate that partitioning or adsorption of polymers may be occurring on apolar polystyrene gel, causing retardation of elution. The 3 μHT and ultrahydrogel columns/DMF solvent system appear to give normal results for MW averages while the 5 μm PLgel columns greatly underestimate the MW. Further work is required to define the observed phenomena in terms of interactions involving solvent, polymer sample type and the type of column packing used. The dimethyl acetamide results indicate that the universal calibration concept is applicable to a variety of types of polymer standards. The trash peaks interfere less with the sample peaks when 3 μHT linear columns are coupled to a Shodex KB 802.5 column.

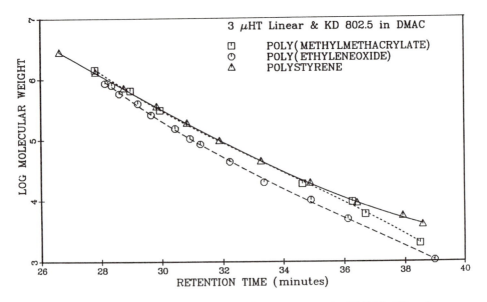

Figure 6.   Primary molecular weight calibration curves for PMMA, PS and PEO in DMAC.

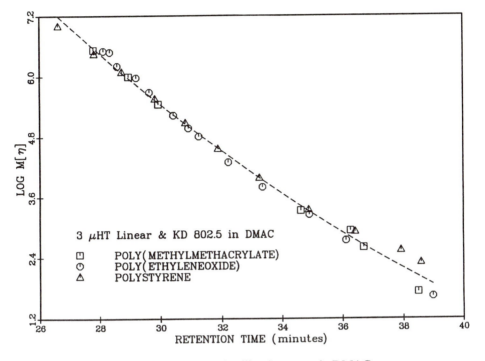

Figure 7. Universal calibration curve in DMAC.

TABLE IV. GPCV/DMAC RESULTS FOR BROAD MWD POLYMERS
(3 μHT LINEAR & SHODEX KD 802.5)

Sample	$\overline{Mn}$ x10⁻³	$\overline{Mw}$ x10⁻³	[η] (dl/g)	K x10⁴	α
Dow 1683					
Nominal Values	100	250	-	1.40[13]	0.68[13]
GPCV/DMAC	81	226	0.67	2.34	0.65
NBS 706					
Nominal Values	136	258	-	1.40[13]	0.68[13]
GPCV/DMAC	106	268	0.69	1.00	0.71

Literature Cited

1.   Haney, M. A. J. Appl. Polym. Sci., 1985, 30, 3037.
2.   Ekmanis, J. Proceedings of International GPC Symposium '89, p.1.
3.   Kuo, C.; Provder, T.; Koehler, M. E. J. Liquid Chromatogr., 1990, 13 (16), 3177.
4.   Malihi, F. B.;Kuo, C.; Koehler, M. E.; Provder, T.; Kah, A. F. In ACS Symposium Series No. 245; Provder, T. Ed., 1984, 281.
5.   Kuo, C.; Provder, T.; Koehler, M. E.; Kah, A. F. In ACS Symposium Series No. 352, Provder, T. Ed., 1987, 130.
6.   Dark, B., Ekmanis, J. L., Harvard, T. J., Huard, T., Lesec, J.; Neilson, R. "The Use of the 150CV GPC/Viscometry System for the Determination of Mark-Houwink Constants K and α", presented at the First International GPC/Viscometry Symposium, Del Lago Resort, Texas, April 24-26, 1991.
7.   Dubin, P. L.; Koontz, S.; Wright, K. L. J. Polym. Sci., Polym. Chem. Ed., 1977, 15, 2047.
8.   Matsuzaki, T.; Inoue, Y.; Ookubo, T.; Moic, S. J. Liq. Chromatogr., 1980, 3, 353.
9.   Tsimpris, C. W.; Suryanarayanan, B.; Maghan, K. G. J. Polym. Sci., Part A-2, 1972, 10, 1837.
10. Ogana, T. J. Liq. Chromatogr., 1990, 13, 51.
11. Hann, N. D. J. Polym. Sci., Polymer Chem. Ed., 1977, 15, 1337.
12. Mourey, T. H.; Bryan, T. G. J. Liq. Chromatogr., 1991, 14, 719.
13. Dawkins, J. V.; Hemming, M. Polymer, 1975, 16, 554.

RECEIVED November 24, 1992

Chapter 16

Absolute M_n Determined by Gel Permeation Chromatography—Differential Viscometry

Judah M. Goldwasser[1]

Mechanics Division, Office of Naval Research, 800 North Quincy Street, Arlington, VA 22217–5000

A new method has been developed to quantitatively determine the number average molecular weight ($\overline{M}_n$) of a polymer sample by GPC using an online differential viscometer as the sole detector. This characterization procedure has been validated by comparing the measured $\overline{M}_n$ of several different polymers and polymer mixtures with expected values. The $\overline{M}_n$ of these samples ranged from 1000 to 1,000,000. The major advantage of this method is that it is not necessary either to utilize the Mark-Houwink constants or to determine the weight fraction of the eluting sample with a concentration detector. A convenient means is thus provided to determine the true $\overline{M}_n$ of copolymers, polymer mixtures, and samples of unknown structure.

Gel permeation chromatography (GPC) has become a very powerful tool in the characterization of polymers. However, the interpretation of a chromatogram to determine the molecular weight distribution parameters (MWDP) of a polymer sample rests primarily upon two factors. These are the conversion of elution volume into molecular weight and the conversion of peak height into weight fraction. This is not generally a problem in the characterization of homopolymers. The introduction of the universal calibration method was a major advance in that it permitted the use of a single set of calibration standards to obtain accurate molecular weight averages for virtually any homopolymer, provided that the Mark-Houwink constants were known *(1-3)*. With the development of the low angle laser light scattering detector (LALLS), absolute molecular weight averages could be obtained even for homopolymers whose Mark-Houwink constants were not available *(4)*.

However, the characterization of copolymers and polymer blends by GPC, especially where there is compositional drift, presents serious obstacles to the determination of accurate molecular weight averages using conventional means. No single set of Mark-Houwink constants can be used to characterize a compositionally heterogeneous polymer sample. In addition, concentration detectors are not generally capable of accurately measuring weight fraction because of their sensitivity to chemical

[1]On detail from Energetic Materials Division, Naval Surface Warfare Center, White Oak Laboratory, Silver Spring, MD 20903–5000

composition as well as concentration. The LALLS detector, which is capable of the direct measurement of molecular weight without employing Mark-Houwink constants and the universal calibration curve, is limited to homopolymers since it requires the use of dn/dc (the change in refractive index with concentration) as a constant in the calculation of molecular weight. Since dn/dc generally changes with chemical composition, LALLS cannot be conveniently used to determine molecular weights of these kinds of samples.

Recently, a differential viscometer was introduced which can be used to determine the intrinsic viscosity of very dilute polymer solutions *(5)*. It was subsequently modified so that it could be used as a GPC detector and determine the intrinsic viscosity of the eluting polymer fractions online *(6)*. With this development, it has become possible to use the universal calibration method to determine $\overline{M}_n$ even where the Mark-Houwink constants are not known or cannot be determined.

Theory

Upon fractionation of a polymer sample by the GPC columns, the specific viscosity η_{sp}, of each eluting species, i , may be obtained from the output of the differential viscometer. In normal operating GPC conditions, the concentration of the eluting species is sufficiently dilute so that the intrinsic viscosity of the eluting species, $[\eta]_i$, may be described according to equation (1),

$$[\eta]_i \cong \frac{\eta_{spi}}{c_i} \tag{1}$$

where c_i, the concentration of the eluting species, is described by equation (2).

$$c_i = \frac{w_i\, C_I\, V_I}{V_S} \tag{2}$$

Here, w_i is the weight fraction of the eluting species, C_I is the injection concentration, V_I is the injection volume, and V_S is the slice volume. C_I and V_I are determined by the operator. V_S is determined by the rate of data collection. Upon substituting equation (2) into equation (1), one obtains

$$[\eta]_i = \frac{\eta_{spi}\, V_S}{w_i\, C_I\, V_I}, \tag{3}$$

and consequently,

$$w_i\, [\eta]_i = \frac{\eta_{spi}\, V_S}{C_I\, V_I}. \tag{4}$$

The intrinsic viscosity of the whole polymer sample, $[\eta]$, may be simply obtained by integrating over the entire range of eluting species as described by equation (5).

$$[\eta] = \sum w_i\, [\eta]_i = \sum \frac{\eta_{spi}\, V_S}{C_I\, V_I} \tag{5}$$

The universal calibration method *(1)* defines the hydrodynamic volume of an eluting polymer species, Vh_i, at infinite dilution, as follows:

$$Vh_i = \frac{4\pi[\eta]_i M_i}{3\phi}, \tag{6}$$

where M_i is the molecular weight of the eluting species and ϕ is Flory's constant. Where one is dealing with a complex polymer sample with heterogeneous composition, several species with different molecular weights but the same molecular size elute simultaneously. In those cases, M_i is really $\overline{M}_{ni}$ *(8)*.

By dividing equation (4) into equation (6), one obtains

$$\frac{4\pi M_i}{3\phi w_i} = \frac{V h_i}{(\frac{\eta_{spi} V_S}{C_I V_I})} . \tag{7}$$

Since

$$\overline{M}_n = \frac{1}{\sum(\frac{w_i}{M_i})} , \tag{8}$$

substitution of equation (7) into equation (8), gives

$$\overline{M}_n = \frac{3\phi C_I V_I}{4\pi V_S \sum(\frac{\eta_{spi}}{V h_i})} . \tag{9}$$

Experimental Section

Characterization of polymer samples was performed using a chromatography system which consisted of a Waters Model 6000A solvent delivery system, a Waters U6K injector, a Molytek Thermalpulse II flowmeter, and a Viscotek Model 100 differential viscometer. In cases where a concentration detector was used, a Waters Model R-401 refractive index detector was employed.

Data were collected using an IBM AT microcomputer which was interfaced to the GPC instrumentation with a Data Translation model DT-2805 A/D board. The software used for both data collection and data reduction was developed in-house using the ASYST v2.0 scientific programming language.

The GPC columns were 10 micron Toyo Soda Micropak H series, 30 cm in length, and with inside diameter of 0.75 cm. For the characterization of polymers with $\overline{M}_n$ in excess of 10,000, four columns were used; one each with packing pore sizes of 1500 Å, 1×10^4 Å, 1×10^5 Å, and 1×10^6 Å. For the characterization of polymers with $\overline{M}_n$ less than 10,000, three 1500 Å columns were used. The eluant was UV grade, unstabilized tetrahydrofuran (THF). The nominal flow rate was 1.0 mL/min.

Sample solutions were delivered using Hamilton 800 series syringes into a sample loop whose volume was less than 75 μL. The vent outlet tube of the injector was turned upward in order to prevent sample run out upon withdrawal of the injection syringe.

The columns were calibrated with narrow distribution Toyo Soda polystyrene standards whose molecular weights ranged from 1,600,000 - 10,000 for the four columns series and from 10,000 - 600 for the three column series.

The polymer samples used in this study were obtained from a variety of sources, including Atlantic Richfield, Aldrich Chemical Company, National Institute for Standards and Technology, Polysciences, and Scientific Polymer Products. All of the samples were characterized by the vendor, and the molecular weight averages were supplied with the product.

Results and Discussion

The intrinsic viscosities of several narrow distribution polystyrene standards were measured according to equation (5) and compared with expected values which were calculated using the Mark-Houwink constants *(9)*. The results are presented in Table I. Agreement between the expected and observed values was within 10% in all cases. The ability to obtain accurate results using the differential viscometer as the only detector relies upon the delivery of samples of known concentration and volume. Concentration is easily determined and controlled by careful sample preparation. However, there were difficulties found in controlling the sample volume, especially when using a manual injector, and this led to inaccurate results. These difficulties were overcome by ensuring that the sample syringe was free of air bubbles, and that no solution ran out of the injector vent tube.

Table I. Comparison of [η] Determined by Differential Viscometry with Expected Values for Narrow Distribution Polystyrene Standards

$\overline{M}_n$	[η] (mL/g)	
	Expected	Observed
1,250,000	296.0	321.0
760,000	205.0	192.0
422,000	126.0	133.0
172,000	69.9	68.5
108,000	43.6	49.9
45,400	26.1	26.6
18,000	13.2	13.6
6,500	6.6	6.5
2,900	4.2	4.9

The precision of the intrinsic viscosity measurements was dependent on whether or not GPC columns were employed. When GPC columns were not connected in line, the precision of measurement to ± 2% could be obtained when replication was made using the same sample solution. When GPC columns were used, the precision of measurement was ± 10%. This is attributed to the differences in peak breadth which is much greater when GPC columns are connected than when they are not. The small errors in baseline choice are insignificant when integrating a sharp, narrow peak, but lead to cumulative errors which become significant when integrating a much broader peak with the same area. Precision of measurement was ± 10% when replication was made using different samples, regardless of whether or not GPC columns were connected in line. This indicates that the major sources of error in this case were cumulative effects of the operations associated with sample preparation and handling.

Measurements of $\overline{M}_n$ were made for several polymers according to equation (9). Figure 1 is a logarithmic plot of measured vs. expected values for several narrow distribution polystyrene and poly(methyl methacrylate) (PMMA) samples with molecular weights ranging from 10^3-10^6, the range of molecular weights for which GPC is generally applied. The slope of the line is 0.947, the intercept is 0.23, and the correlation coefficient, R^2, is 0.995. This illustrates good agreement between the measured and expected values over the entire molecular weight range. $\overline{M}_n$ was also measured for several different broad molecular weight distribution polymers ($\overline{M}_w/\overline{M}_n > 2.5$). The number average molecular weights of these polymers ranged from approximately 1,000-200,000. These results were compared with values supplied by the vendor and are presented in Table II.

Table II. Comparison of $\overline{M}_n$ Determined by GPC/Differential Viscometry with Expected Values for Broad Molecular Weight Distribution Homopolymers

POLYMER	SAMPLE	$\overline{M}_n$ Expected	Measured	$[\eta]$ (mL/g)
Polystyrene	PSA-120	119,500	123,000	83.0
	NBS-706	137,000	133,000	86.0
PMMA	PMA-045	46,400	53,000	31.5
	PMP-065	63,000	67,000	44.3
	PMP-165	165,000	170,000	90.6
Poly(butadiene)	PBP-140	140,000	135,000	219
	PBR-45M	2,800	2,700	15.82
Poly(isoprene)	PIP-010	9,700	9,300	16.26
	PIP-034	33,000	35,000	36.5
	PIP-135	130,000	125,000	96.6
Poly(vinyl acetate)	PVA-090	90,000	86,000	68.5
Poly(caprolactone)	PCP-310	900	1,090	5.56
	PCP-240	2,000	2,180	9.31
	PCP-260	3,000	3,050	12.15
Poly(propylene glycol)	PPG-2000	2,000	2,050	5.47
	PPG-4000	4,000	4,100	10.42

In all cases, the measured values are the average of three replications. Agreement between the measured and expected values were all within 10%. The precision was ± 10-15% and dependent upon the $\overline{M}_n$ of the sample. Precision of measurement was found to be better for polymers with higher $\overline{M}_n$ and lower for those with lower $\overline{M}_n$. There was no relationship found between either accuracy or precision and chemical structure of the polymer sample. This method is completely insensitive to the chemical structure of the polymer being examined since the output of the differential viscometer detector is $w_i[\eta]_i$, and the universal calibration curve is used to determine w_i/M_i. As a consequence, the need to employ parameters which are dependent upon the chemical structure of the polymer sample is eliminated.

The fact that the differential viscometer detector is responsive to $[\eta]_i$ as well as concentration makes it a molecular weight sensitive detector. However, unlike the low angle laser light scattering (LALLS) detector, which has an effective lower molecular weight cutoff of 5,000-10,000, the differential viscometer can be used to obtain molecular weight and intrinsic viscosity information down to the oligomer region. Nonetheless, the detector response is much lower at the low molecular weight end of the peak than the high molecular weight end. As a result, this method is susceptible to errors in the choice of the baseline at the low molecular weight end of the peak. It is known that small errors in choosing the baseline of a GPC peak can exhibit much larger errors in the calculation of $\overline{M}_n$ (7). This was found to be the case here. The values for $\overline{M}_n$ obtained using this method were found to be sensitive to the baseline chosen, and qualitatively, more so for broad distribution samples than for narrow distribution samples. As a result, multiple replication was necessary to ensure good accuracy.

A very important application of the differential viscometer detector is its employment to extend this method to the quantitative characterization of polymer samples with compositional heterogeneity, such as polymer blends and copolymers with compositional drift. In these cases, the chemical composition of the eluting species changes as the polymer elutes upon GPC fractionation. This makes quantitative characterization of such polymer samples impossible using traditional GPC analytical techniques. However, because intrinsic viscosities are additive, the response of the differential viscometer detector is the sum of the products of the weight fractions and intrinsic viscosities of all of the species which elute at any elution volume. In this case, $w_i[\eta]_i = \Sigma w_k[\eta]_k$, where k represents the various species with different chemical composition and possibly different molecular weights, but which all elute simultaneously. Hamielec has shown that the molecular weight average associated with the simultaneous elution of several species at a particular hydrodynamic volume is a local $\overline{M}_n$ (8). Consequently, w_i/M_{ni} is obtained at each elution volume, when characterizing polymer samples which are compositionally heterogeneous. The appropriate summation thus gives an accurate $\overline{M}_n$ for the whole polymer.

This point is illustrated by Figure 2. Several mixtures of polystyrene and poly(tetrahydrofuran) (THF) with varying monomer ratios were examined by both the GPC/viscometry method and by using the traditional concentration detector and calibration curve method. The Mark-Houwink constants used were those for polystyrene. The results are presented in Figure 2. along with the calculated $\overline{M}_n$ line. The values obtained using the GPC/viscometry method are in excellent agreement with the calculated values. This is in contrast to the $\overline{M}_n$ values obtained using the polystyrene calibration curve which show significant deviation from the theoretical line, except where the amount of poly(THF) in the mixture is relatively small (<20%). There are several contributing factors to these deviations. The weight fractions of the eluting species are not correctly determined because the response of the detector is dependent on the chemical structure of the eluting species, and is consequently, not uniform across the GPC curve. In addition, the use of a single set of Mark-Houwink constants is not valid, also because of the presence of more than one polymer in the sample and its changing composition across the GPC curve. These difficulties may be addressed in this simple case of characterizing a binary mixture of known polymers by using two detectors and calibrating their response for each of the components in the mixture so that accurate measurements of the weight fractions of each of the sample constituents can be made. It also requires knowledge of the Mark-Houwink constants for each of the polymers in the mixture. However, it is known that the hydrodynamic volumes of the individual components of polymer blends are influenced by the other components, and consequently, elute at different volumes than would the constituent homopolymer samples (10). This phenomenon compounds the difficulty in attempting to quantitatively determine the $\overline{M}_n$ of even a simple polymer blend. The problem becomes practically intractable for more complicated cases where there are several constituents to the mixture, where polymers of unknown composition, or copolymers with compositional drift are involved. In these cases, it is not possible to accurately determine the weight fractions of the eluting species and their Mark-Houwink constants. In addition, repulsive interactions of the heterodiads markedly modify the hydrodynamic volume of the copolymer(11). The GPC/viscometry

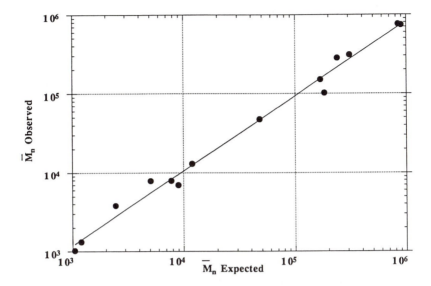

Figure 1. Plot of $\overline{M}_n$ Expected vs. $\overline{M}_n$ Observed for Several Narrow Distribution PS and PMMA Standards

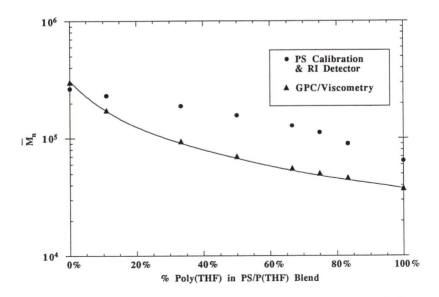

Figure 2. Plot of Calculated Line and Observed Values of $\overline{M}_n$ vs. % Poly(THF) in Blends of Polystyrene with Poly(THF)

method, by contrast, accounts for these complexities inherently and consequently may be used to quantitatively characterize complex polymer samples as easily as simple homopolymers.

Several different polymer mixtures with a broad range of molecular weights were also examined by GPC/viscometry. In the high molecular weight region, $\overline{M}_n$ and [η] were determined for 1:1 mixtures of narrow distribution polystyrene and PMMA with varying molecular weights. The results were compared with calculated values and are presented in Table III.

Table III. Comparison of $\overline{M}_n$ and [η] Determined by GPC/Differential Viscometry with Expected Values for 1:1 Mixtures of Narrow Distribution Polystyrene and PMMA Standards

$\overline{M}_n$ of Homopolymers		$\overline{M}_n$ of Mixture		[η] (mL/g)	
PS	PMMA	Expected	Measured	Expected	Measured
900,000	400,000	553,900	522,000	137	122
900,000	240,000	379,000	377,000	124	116
310,000	845,000	453,000	423,000	112	104
170,000	845,000	283,000	265,000	97.8	94.0
48,000	845,000	90,800	91,600	77.2	72.1

Mixtures of a number of different broad distribution polymer standards which were available in a lower molecular weight region were examined. These polymers included polystyrene (PS), poly(methyl methacrylate (PMMA), poly(vinyl acetate) (PVAC), and poly(butadiene) (PBD). $\overline{M}_n$ and [η] values were determined for these mixtures with various composition ratios and compared with the calculated values as shown in Table IV.

Table IV. Comparison of $\overline{M}_n$ and [η] Determined by GPC/Differential Viscometry with Expected Values for Various Mixtures of Broad Distribution Polymers

Ratio	Polymers	$\overline{M}_n$ of Mixture		[η] (mL/g)	
		Expected	Measured	Expected	Measured
7:3	PS:PMMA	85,000	87,000	72.3	68.1
3:7	PS:PMMA	57,000	60,000	48.0	49.0
1:1	PS:PVAC	108,000	112,000	79.1	78.7
1:1	PMMA:PVAC	61,000	63,000	49.6	48.4
1:1	PS:PBD	137,000	144,000	157	148

As in the homopolymer case, the reported values for $\overline{M}_n$ and [η] of these mixtures are the average of three replicated measurements to ensure good accuracy. Agreement with calculated values was within 7% in all cases, and precision was ± 10%.

These results are a good indication of the robustness of the GPC/viscometry method. Accurate measurements of $\overline{M}_n$ and [η] can be conveniently be obtained for polymers across a wide range of molecular weights and involving several chemical structures. Insensitivity to chemical structure has been demonstrated for several samples with different polymer structures. This is true for samples which are compositionally heterogeneous as well as for homopolymers, and for polymer samples with broad molecular weight distributions as well as narrow distributions.

The factors which are give rise to these advantages, however, are also responsible for the limitations of this method. The sensitivity of the differential viscometer to molecular weight results in less sensitivity at the the low molecular weight side of the GPC peak. As a result, the $\overline{M}_n$ is very sensitive to baseline choice at the low molecular weight side of the peak, and care must be taken to minimize error. At present, it is also recommended that several replications be be and averaged in order to ensure good accuracy. Steps may be taken to enhance the signal-to-noise ratio by optimizing the injection concentration and volume. This may also be accomplished by increasing the solvent flow rate, with some sacrifice of resolution. [η] measurements, on the other hand, are not very sensitive to baseline choice.

Conclusions

A new method for the quantitative determination of $\overline{M}_n$ by GPC has been developed. This method requires only the universal calibration curve and the differential viscometer detector. The Mark-Houwink constants and a concentration detector are not required. This method may be employed to characterize polymer samples of unknown composition, as well as samples with compositional drift such as polymer blends and copolymers. Because the viscometer is a molecular weight sensitive detector, results obtained using this procedure are sensitive to the baseline choice at the low molecular weight side of the chromatogram.

Acknowledgments

This work was supported by the Office of Naval Research and the Independent Research Program of the Naval Surface Warfare Center. The author is extremely grateful to Dr. Wallace Yau and Dr. Howard Barth of DuPont, Inc. for invaluable discussions and to Peggy Ware and Geoffery Smith also of DuPont, Inc. for providing important technical support.

References

1. Grubisic, Z.; Rempp, P.; Benoît, H. *J. Polym. Sci., Polym. Phys. Ed.*, **1967**, *5*, 753.
2. Nichols, E. *ACS Advances in Chemistry Series*, **1973**, *125*, 148.
3. Rudin, A.; and Wagner, R. A. *J. Appl. Polym. Sci.*, **1976**, *20*, 1493.
4. Jordan, R. C. *J. Liq. Chromatogr.*, **1980**, *3*, 439.
5. Haney, M. A. *J. Appl. Polym. Sci.*, **1985**, *30*, 3023.
6. Haney, M. A. *J. Appl. Polym. Sci.*, **1985**, *30*, 3037.
7. Tchir, W. J.; Rudin, A.; Fyfe, C. A. *J. Polym. Sci., Polym. Phys. Ed.*, **1982**, *20*, 1443.
8. Hamielec, A. E.; Ouano, A. C. *J. Liq. Chromatogr.*, **1978**, *1*, 111
9. Spatorico, A. L.; Coulter, B. *J. Polym. Sci., Polym. Phys. Ed.*, **1973**, *11*, 1139.
10. Kok, C. M.; Rudin, A. *Makromol. Chem.*, **1981**, *182*, 2801
11. A. Dondos, A.; Rempp, P.; Benoît, H. *Makromol. Chem.*, **1974**, *175*, 1639.

RECEIVED October 9, 1992

Size-Exclusion Chromatography: High-Temperature, Ionic, and Natural Polymer Applications

Chapter 17

Size-Exclusion Chromatographic Assessment of Long-Chain Branch Frequency in Polyethylenes

Simon Pang and Alfred Rudin

Institute for Polymer Research, Department of Chemistry, University of Waterloo, Waterloo, Ontario N2L 3G1, Canada

Long chain branch frequencies in various polyethylenes have been characterized by SEC using viscometer and light scattering detectors. The SEC estimations involve use of a Zimm-Stockmayer relation between an assumed molecular structure and radius of gyration of the macromolecule. Here, we have compared such estimates of branch frequency with values measured by ^{13}C analyses. Because of experimental noise some SEC analyses may indicate the presence of long branching where none is in fact present. For high pressure process, low density polyethylenes, the SEC estimates are in reasonable coincidence with ^{13}C NMR results, which are believed to be the referee method. We conclude that while SEC measurements of long chain branch frequency cannot be assumed a priori be to very accurate, they rank various polyethylenes correctly and probably provide long branch content values that are within a factor of 2 of the "true" values.

It has been recognized that long chain branching may have important effects on the properties of polymers in which this feature may occur. Attempts to quantify long branch concentrations have been hampered by analytical difficulties and uncertainties in the assumptions of relations between molecular weight, long branch frequency and hydrodynamic volume, which is the basic variable in SEC separations (1).

SEC is the only current analytical technique that can provide information on the variation of long branching with molecular weight, without fractionation of the polymer. This is possible, of course, only with the use of detectors that can measure molecular weights directly.

At equal SEC retention time and infinite dilution, the molecular weights of linear species and long branched versions of the same polymer are related by:

$$[\eta]_b M_b = [\eta]^* M^* \tag{1}$$

0097–6156/93/0521–0254$06.00/0

where the subscript b and superscript * refer to branched and linear macromolecules that have the same SEC elution volume. Now consider a branched and linear version of the same polymer, both with the same molecular weight. In that case the intrinsic viscosity $[\eta]_l$ of the linear polymer will be greater than that of the branched species, $[\eta]_b$ in the SEC solvent. The ratio of the intrinsic viscosities is:

$$g' = \frac{[\eta]_b}{[\eta]_l} \tag{2}$$

(Note that $M_b = M_l \geq M^*$)
For monodisperse versions of the linear polymer in the SEC solvent the Mark Houwink relation is:

$$[\eta] = KM^a \tag{3}$$

From equations 1 and 3:

$$g' = \left[\frac{M^*}{M_b}\right]^{a+1} \tag{4}$$

In order to relate g' to actual molecular size it is necessary to consider the ratio of radii of gyration of the branched and linear polymers with molecular weights $M_l = M_b$. That is:

$$g = \frac{\langle R_G^2 \rangle_b}{\langle R_G^2 \rangle_l} < 1 \tag{5}$$

Various relations have been proposed (2,3) of the form:

$$g' = g^k \tag{6}$$

with k theoretically between 0.5 and 1.5. More recent experiments have shown that k may not be the same for polyethylenes with different long branch concentrations (4) and hence may vary through the molecular weight distribution of a single polymer sample.
Equations 4 and 6 provide:

$$g' = \left[\frac{M^*}{M_b}\right]^{a+1} = g^k \tag{7}$$

Equation 7 is frequently used to measure long chain branching frequency in polyethylene (5,6). At any given retention time, M_b is measured directly with a low angle laser light scattering detector (LALLS) or a continuous viscometer (CV), while M^* is calculated from the universal calibration curve for the linear polymer of the same type as the branched material. The long branch frequency is expected to be reflected in the value of the experimental parameter g'.
Several assumptions must be invoked in order to estimate the actual number

of long branches per molecule. In particular, one must assume a branch structure for the macromolecule and a value for the exponent k in equation 7. This paper examines such assumptions explicitly and compares SEC estimates of long chain branch frequency with those from an independent assessment with ^{13}C NMR.

Most SEC calculations assume that the polymer is randomly branched and contains trifunctional branching points. Then g (equation 7) can be related to the weight average number of branch points, n_w, per molecule according to Zimm-Stockmayer relation (7):

$$g = \frac{6}{n_w} \left\{ \frac{1}{2} \frac{(2+n_w)^{0.5}}{n_w^{0.5}} \ln \left[\frac{(2+n_w)^{0.5} + n_w^{0.5}}{(2+n_w)^{0.5} - n_w^{0.5}} \right] -1 \right\} \qquad (8)$$

Finally, the long chain branching frequency λ per 1000 carbon atoms can be calculated using the following equations:

$$\lambda = \frac{n_w(14)(1000)}{M_b} \qquad (9)$$

where M_b is the molecular weight measured for the branched species at any given elution volume.

Application of equation (8) to low density polyethylenes is somewhat controversial. The assumption that branch points are random and not clustered is debatable and ^{13}C NMR analyses indicate a variable number of tetrafunctional as well as trifunctional branch junctions. In addition, the theoretical relation was derived for a Theta solution, whereas SEC analyses are performed in good solvents.

Studies of copolymers of ethylene with 1-olefins have shown that C_6 branches are not registered as long in SEC analyses, whereas C_{12} branches are measured as long. The minimum branch length for long branching is therefore between C_6 and C_{12} (4) at least for copolymers in which the 1-olefin concentration is less than 7 mole per cent. Present-day ^{13}C NMR analyses usually measure polyethylene branches six carbons or larger as long. This is about the same length that is apparently measured by SEC. SEC analyses measure long chain branching as a function of molecular weight while ^{13}C NMR analyses "see" the whole polymer. Here, we have averaged SEC estimates over the molecular weight distribution in order to compare the SEC-Zimm/Stockmayer estimates with the absolute ^{13}C NMR analyses.

Experimental Details

Instruments.(8) The SEC system used in this study consisted of a high temperature chromatograph equipped with a differential refractive index (DRI), LDC/Milton Roy KMX-6 low angle laser light scattering (LALLS), Viscotek Model 100 continuous differential viscometer (CV) detector, an Erma Optical Works Ltd. ERC-3510 on-line degasser, a Molytek thermopulse flowmeter, and a set of Jordi columns which comprised a mixed bed column and a 1000Å linear column. The experiments were run with a flow rate 0.7 mL/min at 145°C.

The LALLS photometer with a high temperature flow through cell was serially connected with the column. Scattering intensity data were collected using a 6-7° annulus with a 6328Å wavelength, He-Ne laser. The DRI and CV detectors were connected in parallel to the LALLS detector. The ratio of the flow volumes between the DRI and CV lines was approximately 50:50. A flowmeter was connected in series with the DRI to monitor the instantaneous flow rate between the branches during the experimental runs. The apparatus is sketched in Figure 1. Polymer concentration in the eluent was monitored with the DRI detector. The polymer concentration in the injected sample was 2-2.5 mg/ml. The mobile phase was filtered through an on-line 0.5 μm tetrafluoroethylene filter just before the LALLS cell.

The value of (dn/dc) for the polyethylenes were determined independently with a LDC/Milton Roy KMX-16 differential refractometer. This value was found to be 0.091 cm^3/g.

Materials and Sample Preparation. All solutions for analysis were prepared in filtered 1,2,4-trichlorobenzene (TCB), the same solvent used as the SEC eluent. Polymer solutions were prepared by dissolving known quantities of polyethylenes and diluting to the desired volume with the filtered TCB solvent. Dissolution of PE samples were achieved by rotating the samples at 160°C for 16 to 24 hours. To prevent oxidative degradation of LDPE, 0.1 weight-percent of antioxidant (Irganox 1010) was added. The column set was calibrated using 25 polystyrene standard samples with molecular weights ranging from 950 to 15,000,000.

Complete dissolution of the polymers were assumed to have been achieved when the LALLS detector trace was free of spikes (*9*). Higher molecular weight linear polyethylenes may require longer dissolution times than those used in this study. Noise in the signals was not suppressed and the calculated molecular weight distribution curves were not smoothed.

^{13}C NMR Analyses of Long Chain Branch Frequency. ^{13}C NMR spectra of the polyethylenes were obtained with a Bruker AM-300 spectrometer operating at 75.5 MHz and equipped with a Bruker Aspect 3000 computer and a B-VT 1000 variable temperature unit. All samples were prepared in 10 mm o.d. tubes as 40% (w/w) concentrations in 1,2,4-trichlorobenzene solvent, and were run at 125°C. An inverse-gated pulse sequence was used (to minimize nuclear Overhauser effects) with a 12 second relaxation delay and 190 degree pulse width. 32768 data points were collected over a 140 ppm sweep with an acquisition time of 1.54 seconds. No deuterium lock was used in the samples, since field drift over data collection time was considerably less than the line widths obtained. Chemical shifts were referenced internally to the major backbone methylene carbon signal at 29.99 ppm. Peak areas were determined by planimetry, which is believed to be more accurate than instrumental integration.

Six carbon branches are detected at 32.22 ppm. True long chain branches can be seen at 32.18 ppm. This signal is evident in our analyses if the long branch frequency is greater than about 10 percent of the six carbon branch frequency. The overlap of the two signals is an analytical problem only with ethylene-octene

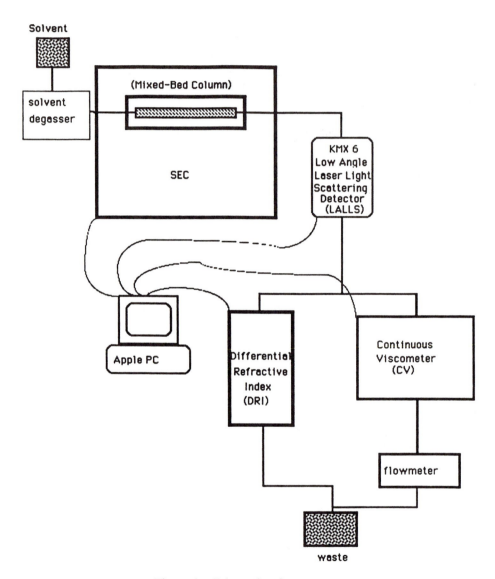

Figure 1. Schematic of apparatus.

copolymers. In that case, NMR analyses were performed under the same conditions as above, except that 5 mm o.d. tubes were used and the polymer concentration was reduced to 10% (w/w) in trichlorobenzene.

Results and Discussion

Figure 2 shows the raw SEC chromatograms of NBS SRM 1476 from the DRI, CV and LALLS detectors. This is a low density, high pressure process polyethylene. Figure 3 shows the molecular weight vs. elution volume plot of this LDPE estimated from the universal calibration, continuous viscometer and LALLS analyses. The cut-off lines present in the plot were determined by selecting the regions where the deviations (apparent oscillations in the trace) of the calculated molecular weight from LALLS or the intrinsic viscosity from CV exceeded preset threshold levels ($\pm 3\%$ of the LALLS peak height and $\pm 1\%$ for the case of the CV analysis). In general, the cut-off at the low elution volume tail (high molecular weight) is governed by the strength of the DRI signals which are used as concentration data. At the high elution volume end, the cut-off is dictated by the noise in the LALLS or CV signal. Figure 4 shows the molecular weight-solution volume plots of a linear polyethylene (NBS 1475); Figure 5 summarizes the data for an ethylene-octene copolymer (LLDPE-A); Figure 6 is that of an ethylene-butene copolymer (LLDPE-B) and Figure 7 shows molecular weight-solution volume data for a high pressure process low density polyethylene (LDPE-A). It is clear that LDPE molecular weights estimated from universal calibration are lower than those obtained from CV and LALLS analyses. Figures 8-12 show calculated long chain branch frequencies against molecular weight plots for various polyethylenes. (All these calculations were made with $k=0.7$ in Eq. (7)). These observations for LDPE's are in fairly good agreement with the CV study of Mirabella and Wild (*10*) and an earlier LALLS study in our laboratory (*6*).

All data are presented here without noise suppression or curve smoothing. The discrepancies in the low molecular weight region reflect noise in the LALLS signal. By contrast, the LALLS is more sensitive than the CV in the very high molecular weight region.

Table I lists branch character for the various polyethylenes, from ^{13}C NMR analyses. The LLDPE's are all free of long branches. The LDPE's contain the expected branch types: ethyl at tertiary and quaternary carbon atoms, butyl, amyl and long branches.

Table II shows molecular weight and long branch frequency averages from SEC/CV and SEC/LALLS analyses. The LCB-molecular weight data were averaged across the measured molecular weight range between the cut-off points of the various detectors. The long branch frequency from ^{13}C NMR analyses of the whole sample is also reported.

The NMR branch frequency data are considered to be precise to within about ± 10 percent. The NMR technique is sensitive to the whole sample, while the SEC data are clipped at the low and high molecular weight ends of the distribution, as noted above. The SEC estimates of long chain branch frequency are certainly not more sensitive nor precise than the NMR values.

The SEC technique has indicated the presence of low levels of long branching

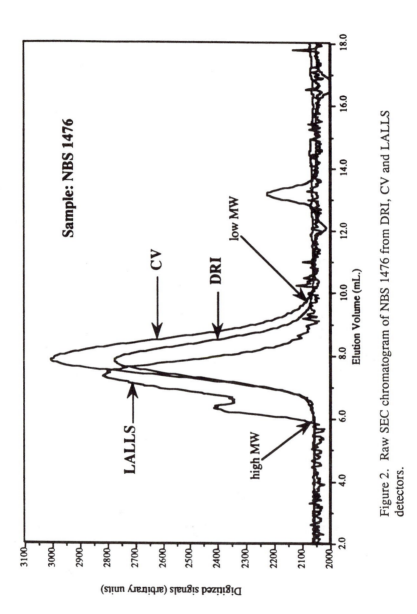

Figure 2.   Raw SEC chromatogram of NBS 1476 from DRI, CV and LALLS detectors.

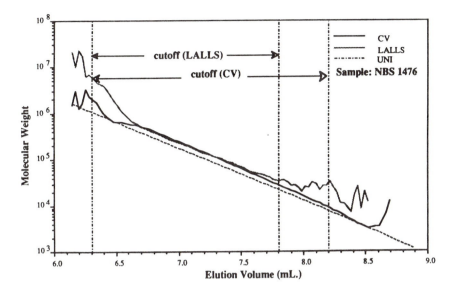

Figure 3. Molecular weight vs. elution volume plot of NBS 1476 LDPE from DRI-universal calibration, CV and LALLS analyses.

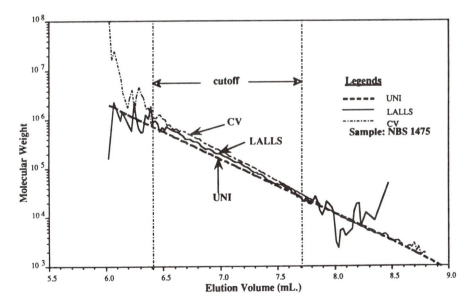

Figure 4. Molecular weight vs. elution volume plot of NBS 1475 linear PE from DRI-universal calibration, CV and LALLS analyses.

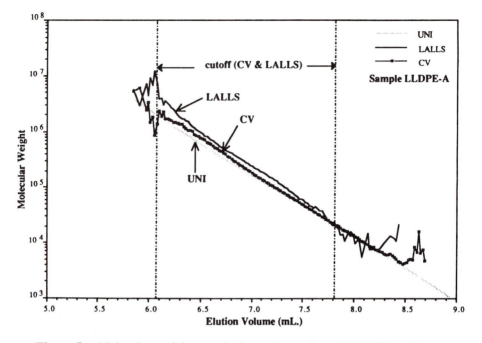

Figure 5. Molecular weight vs. elution volume plot of LLDPE A (slurry process, ethylene-octene copolymer) from DRI-universal calibration, CV and LALLS analyses.

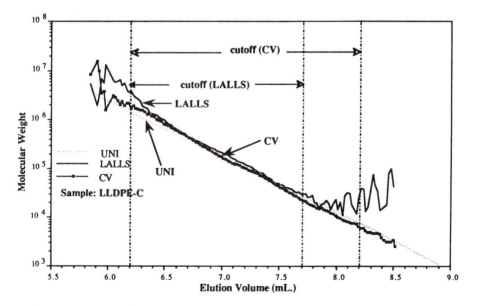

Figure 6. Molecular weight vs. elution volume plot of LLDPE C (gas phase process, ethylene-butene copolymer) from DRI-universal calibration, CV and LALLS analyses.

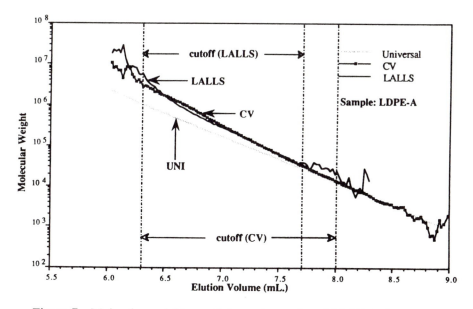

Figure 7. Molecular weight vs. elution volume plot of LDPE A from DRI-universal calibration, CV and LALLS analyses.

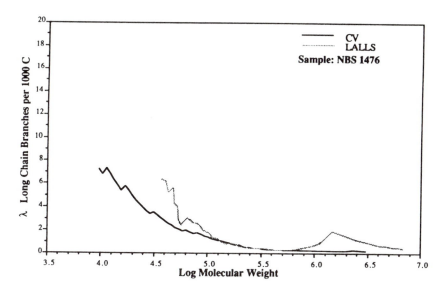

Figure 8. Long chain branch frequency distribution plot of NBS 1476 with k = 0.7, in equation 7 from both CV and LALLS data.

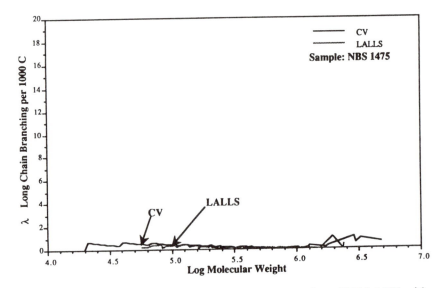

Figure 9. Long chain branch frequency distribution plot of NBS 1475 with k = 0.7, in equation 6 from both CV and LALLS data.

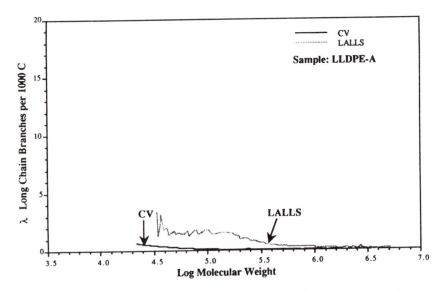

Figure 10. Long chain branch frequency distribution plot of broad distribution sample LLDPE A with k = 0.7, in equation 6 from both CV and LALLS data.

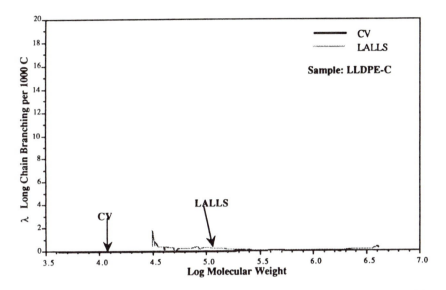

Figure 11. Long chain branch frequency distribution plot of LLDPE C with k = 0.7, in equation 6 from both CV and LALLS data.

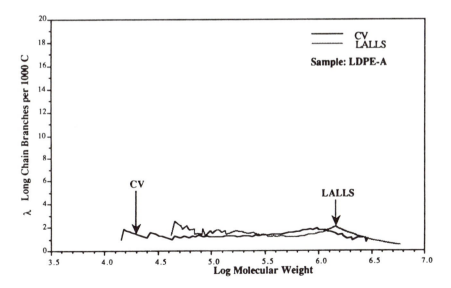

Figure 12. Long chain branch frequency distribution plot of broad LDPE A with k = 0.7, in equation 6 from both CV and LALLS data.

Table I. Long Chain Branching Concentrations from ^{13}C NMR Analysis

Sample	Type of Branching	ppm	Branches/1000C
LLDPE - A	LCB(>=6C)	32.20	*0
LLDPE - B	LCB(>=6C)	32.20	0
LLDPE - C	LCB(>=6C)	32.20	0
LDPE - A	amyl ($3B_5$) LCB (>=6C) butyl ($2B_4$) ethyl ($1B_2$) ethyl ($1B_2'$)	32.70 32.20 23.27 11.05 8.15	2.59 4.54 9.73 2.76 0.52
LDPE - B	amyl ($3B_5$) LCB (>=6C) butyl ($2B_4$)	32.67 32.20 23.37	2.55 3.80 8.67
NBS 1476	amyl ($3B_5$) LCB (>=6C) butyl ($2B_4$) ethyl ($1B_2$) ethyl ($1B_2'$)	32.67 32.20 23.37 11.20 8.20	2.54 2.53 11.16 1.53 0.51
NBS 1475	LCB (>=6C)	32.20	0

*8.13 hexyl branches per 1000 carbons; no long branches were detected.

Table II. Molecular Weight and Long Branch Frequency Averages from SEC/CV and SEC/LALLS Analyses

Sample	LALLS			CV			Intrinsic Viscosity dl/g (b)	LALLS LCB/1000C	CV LCB/1000C	^{13}C NMR LCB/1000C
	$\overline{M}n$	$\overline{M}w$	$\overline{M}z$	$\overline{M}n$ (a)	$\overline{M}w$	$\overline{M}z$				
NBS 1476	39,600	126,500 (128,700)	1,091,500	32,900 (35,200)	116,400	796,600	0.904 (0.909)	2.01	1.82	2.5
NBS 1475	20,000	54,230 (54,300)	208,080	19,290 (19,320)	54,340	165,780	1.234 (1.247)	0.02	0.01	0.0
LLDPE-A	46,500	194,090 (186,160)	873,420	34,240 (32,260)	141,030	513,800	1.629 (1.649)	0.28	0.08	0.0
LLDPE-B	46,620	201,640 (195,260)	866,740	35,350 (33,220)	152,330	590,130	1.712 (1.734)	0.25	0.07	0.0
LLDPE-C	55,040	171,200 (170,490)	872,110	27,930 (29,920)	138,070	790,710	1.761 (1.781)	0.00	0.00	0.0
LDPE-A	51,820	237,760 (220,950)	2,079,190	18,570 (20,400)	225,430	1,113,120	0.944 (0.949)	2.07	1.87	4.5
LDPE-B	73,730	345,730 (341,760)	2,004,040	35,450	348,200	1,864,560	1.113 (1.119)	2.58	2.45	3.8

(a) values in parentheses estimated from using the area under the chromatogram only with Goldwasser method (*11*)
(b) values in parentheses estimated from using the area under the chromatogram only, with the known mass of polymer injected.

in three samples (NBS 1475, LLDPE's A and B) where none was detected by ^{13}C NMR spectroscopy. The discrepancies between average SEC estimates and NMR values are greater for the LALLS than the CV detector because of the higher noise level in the former.

The NMR values are consistent with the results of SEC analyses for most samples except for those noted. The differences seen reflect detector noise in the molecular weight regions which cannot be analyzed by SEC at present.

In branched polymers, the SEC estimates of long branch frequency are lower than the NMR values, which can be taken to be more nearly correct, since they are not subject to any assumptions regarding the relation between molecular structure and radius of gyration. Another uncertainty is in the assigned value of the exponent k in equation 7. Other values of k would bring the SEC and NMR data somewhat closer. We have not tried to select an optimum magnitude of this exponent because no single value would make the two sets of data coincide, and because there is evidence that a single k may in fact not apply to all polyethylenes (12,13).

Mirabella and Wild (10) report that the frequently observed increase in LCB in the low molecular weight region of the chromatogram is probably fallacious because of the insensitivity of the molecular weight detectors to low molecular weight species. Our data are not in disagreement with these authors. Our LCB-molecular weight values for LDPEA and LDPEB (Figures 12 and 13) show no significant variation with molecular weight. NBS SRM 1476 plots have similar shapes in both studies, although the branch frequencies estimated from SEC and NMR analyses in our work are about double those reported by Mirabella and Wild (10). The reasons for these discrepancies are not clear.

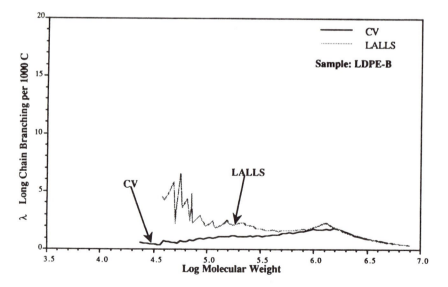

Figure 13. Long chain branch frequency distribution plot of LDPE B with k = 0.7, in equation 6 from both CV and LALLS data.

It appears in summary then that SEC analyses of long branch frequency that use the Zimm-Stockmayer randomly branched molecule as a model provide reasonable results. The coincidence of SEC and NMR average values could have been improved by an arbitrary increase of the exponent k in the theoretical range of this parameter, but this would not be a constant value for all polymers.

Acknowledgment

This research was supported by the Natural Sciences and Engineering Research Council of Canada.

References

1.   Rudin, A. In *Modern Methods of Polymer Characterization*, Barth, H. and Mays, J., Eds.; Wiley, New York, 1991, p. 103.
2.   Zimm, B.H.; Kilb, R.W. *J. Polym. Sci.* **1959**, *37*, 19.
3.   Berry, G.C. *J. Polym. Sci.* **1971**, *9*, A2, 687.
4.   Grinsphun, V.; Rudin, A.; Russell, K.E.; Scammell, M.V. *J. Polym. Sci., Polym. Lett. Ed.* **1986**, *24*, 1171.
5.   Axelson, D.E.; Knapp, W.C. *J. Appl. Polym. Sci.* **1980**, *25*, 119.
6.   Rudin, A.; Grinsphun, V.; O'Driscoll, K.F. *J. Liq. Chromatogr.* **1984**, *7*, 1809.
7.   Zimm, B.H.; Stockmayer, W.H. *J. Chem. Phys.* **1949**, *17*, 1301.
8.   Pang, S.; Rudin, A. *Polymer (London)* **1991** accepted.
9.   Grinsphun, V.; O'Driscoll, K.F.; Rudin, A. In *ACS Adv. Chem. Ser.*, Provder, T., Ed., *ACS Adv. Chem. Ser.*, 245, 273, (1984).
10.   Mirabella Jr., F.M.; Wild, L., *ACS Adv. Chem. Ser.*, 227, C.D. Craver, and T. Provder, eds., 23 (1990).
11.   Goldwasser, J.M. *Intl. GPC Sympos.* **1989**, p. 150, Waters Chromatogr. Div., Newton, MA.
12.   Grinsphun, V., Rudin, A., Russel, K.E. and Scammell, M.V., *J. Polym. Sci., Polym. Lett. Ed.*, 24, 1171 (1986).
13.   Bugada, D.C., Rudin, A., *Eur. Polym. J.*, 11, 847 (1987).

RECEIVED October 7, 1992

Chapter 18

Gel Permeation Chromatography—Fourier Transform IR Spectroscopy To Characterize Ethylene-Based Polyolefin Copolymers

R. P. Markovich[1], L. G. Hazlitt[1], and Linley Smith-Courtney[2]

[1]Polyolefins Research, Dow Chemical USA, Freeport, TX 77541
[2]Division of Mathematical and Information Sciences, Sam Houston State University, Huntsville, TX 77340

A Fourier transform infrared (FT-IR) spectrometer has been coupled with a high temperature gel permeation chromatography (GPC) instrument to provide a powerful tool for the characterization of ethylene based polyolefin copolymers. The combination of these devices provided the ability to simultaneously characterize the molecular weight (MW), the molecular weight distribution (MWD), and chemical composition (such as, the comonomer or branching concentration) as a function of molecular weight. Unique problems and solutions associated with development of GPC/FT-IR system for ethylene based copolymers will be presented.

GPC is a powerful analytical tool frequently used for characterization of polymers. GPC separates the polymer molecules according to size. The size of any polymer molecule is dependent on its molecular weight, composition, branching and microstructure. With appropriate calibration between molecular weight and molecular size, GPC can provide information about the molecular weight and the molecular weight distribution of the whole polymer.(1) The molecular weight and molecular weight distribution data are used to correlate with polymer properties, especially polymer physical properties.

IR spectroscopy is also a powerful tool frequently used for characterization of polymers. The IR spectra can provide quantitative and qualitative information about IR active functional groups in polymers and therefore can indirectly provide information about the microstructure of polymers (including information about branching and comonomer concentration). This information can be provided almost instantaneously by the use of an FT-IR because it is capable of obtaining a IR spectrum from 4000-700 cm^{-1} in less than 1 second. Use of an FT-IR as a detector for GPC provides the ability to simultaneously characterize the molecular weight, the molecular weight distribution and to identify and quantify IR active functional groups. One thus has the ability to characterize the concentration of IR active functional groups as a function of molecular weight. GPC/FT-IR has been used to provide such information for different applications and different materials(2) including high density polyethylene (HDPE) and linear low density polyethylene (LLDPE) (3,4). In this paper, unique problems and solutions associated with development of GPC/FT-IR for ethylene based copolymers will be presented. Further, examples of the characterization information obtained via this method are presented.

0097–6156/93/0521–0270$06.00/0

Experimental Details

Materials. Experiments were conducted with the following ethylene based copolymer types: ultra low density polyethylene (ULDPE), linear low density polyethylene (LLDPE), ethylene acrylic acid (EAA) and ethylene carbon monoxide (ECO) copolymers.

Conditions for GPC/FT-IR Analysis. Figure 1 illustrates the key features of the GPC/FT-IR setup. A Waters 150C ALC/GPC was connected via a heated transfer line to a BioRad FTS-60 FT-IR. The Gel Permeation Chromatography operating conditions were: column = Polymer Laboratories Gel Mixed Bed 20 Micron; operating temperature = 140°C; mobile phase = 1,2,4-Trichlorobenzene; Flow rate = 1.0 ml/min; sample concentration 0.25 g/50ml; injection size = 400 µl; Run time = 60 minutes; mass detectors = FT-IR and differential refractometer; molecular weight standards = a combination of the polystyrene standards ranging in M_W from 600 to 8,400,000. Column calibration for molecular weight estimation was achieved separately (column effluent directed to the differential refractive index mass detector) using the method described by Williams(5) for use with linear polyethylene systems. During the actual analysis the column effluent was directed exclusively to the spectrometer with appropriate adjustments for elution elution volume to compensate for detector offset. Heated transfer lines were maintained at 140°C. The FT-IR optical bench operating conditions were: sample cell = heated, stainless steel flow thru cell with zinc selenide windows; sample cell operating temperature = 140°C; cell pathlength = 1 mm; spectrum type = single beam; scan number = 50 (coadded); scan frequency = 20,000 Hz; Resolution = 8 cm^{-1}; aperture = open; detector = wide band, mercury cadmium telluride (MCT); data station = Model 3260; 40 spectra were saved per GPC analysis; up to eight samples could be analyzed unattended in a run.

The Bio-Rad FTS-60 data station was connected via RS232 cable to an IBM AT clone running GPC data acquisition software. Single beam spectra were taken in real time at 30 second intervals by commands sent from the PC to the FT-IR data station. The single beam spectra were stored on the FT-IR data station hard disk. Forty single beam spectrum were taken per sample. Up to eight samples were analyzed per run.

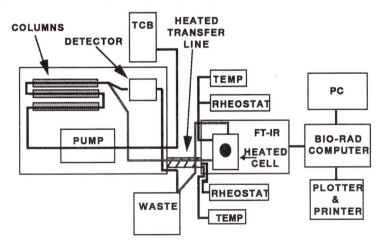

Fig. 1. GPC/FT-IR instrumental set-up.

Data Processing and Analysis. After a GPC/FT-IR run was completed, data from the single beam spectrum stored on the FT-IR data station hard disk were transferred to the PC for further processing and analysis. For each single beam spectra, only the single beam spectral intensities for a selected set of wavelengths were transferred to the PC. This was done to provide sufficient information for the features of interest and at the same time minimize both memory requirements and data processing time. The wavelengths that were used are described below in Table 1 along with the features associated with them. Software programs on the PC processed the data and generated normalized molecular weight and chemical composition distribution curves. Some of the software was developed in-house specifically for this application.

Table 1

Description of Absorbance	Wavenumber (cm^{-1})
Carbonyl	1706, 1714
Methylene Symmetrical Stretching	2846, 2854, 2862
Methylene Asymmetrical Stretching	2918, 2826, 2934
Methyls	2950, 2958, 2966

Background Intensity Baseline Method for Calculation of Solute Absorbance. It was necessary to develop a non-classical method to calculate solute absorbance. Determination of absorbance values via the classic method of using the ratio of a background or reference single beam spectra (radiant intensity transmitted through solvent and sample cell) from a sample single beam spectra (radiant intensity transmitted through solute, solvent and sample cell) was not possible because of the following factors: low concentrations of sample, baseline drift and significant loss of energy due to the combined effect of the solvent, the cell pathlength, and the zinc selenide IR cell windows.

The unique feature of the method was the use of background intensity baseline functions, in place of a discrete background single beam spectra, to determine background (or reference) intensities. The background intensity baseline is a line equation relating background intensity to elution volume. A unique background intensity baseline was determined for each of the wavenumbers used during an analysis. The line equation was determined by calculating the least-squares best line fit between the first three and last three single beam intensity values obtained at that wavenumber during a GPC/FT-IR analysis. An example of background intensity baseline is illustrated in Figure 2 for 2926 cm^{-1}.

The absorbance of the solute at any one of the forty sample points for a given wavenumber was calculated using the following derivation of Beer's law:

$$A_{ij} = \log \frac{B_{ij}}{S_{ij}}$$

where i is the wavenumber, j is the elution volume (or sample point), B is the background intensity at the given wavenumber and elution volume which was calculated from the background intensity baseline for that wavenumber and S is the single beam intensity value at the corresponding wavenumber and elution volume. In Figure 2, the corresponding background intensity (B) and sample single beam intensity (S) values which would be used to calculate the solute absorbance at wavenumber 2926 cm-1 and elution volume 21.25 cc are highlighted for illustration.

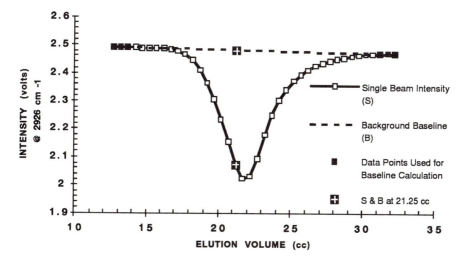

Fig. 2. Background intensity baseline method.

Mass Detection and Determination of Weight Percent Versus Elution Volume. The FT-IR was used as the mass detector. All eleven of the wavenumbers listed in Table 1 were used for mass detection. Since spectra were taken at 40 equally spaced times during the GPC analysis; we were able to determine the weight percent of polymer at each these times. The weight % of polymer was determined using the following equation:

$$\text{Weight \%} = 100 \times (A_1 / A_{Total})$$

where A_1 is the sum of the absorbance values for the eleven wavenumbers for a specific spectra and A_{Total} is sum of the absorbance values for the eleven wavenumbers for all 40 of the spectra obtained during a GPC analysis. Since the spectra were taken at known times during the GPC analysis, the weight % data could be reported as a function of time, elution volume or molecular weight.

Quantitative Determination of Methyls and Olefin Copolymer Concentration. Standards ranging in methyl concentrations from 1 to 100 methyls per 1000C carbons (M/1000C) were used to define a calibration curve. The standards were NBS 1483 (0.97 M/1000C), NBS 1482 (2.46 M/1000C), Dotriacontane (62.5 M/1000C), and Eicosane (100.0 M/1000C). The calibration curve related M/1000C to the ratio of methyl absorbance divided by total absorbance. The methyls absorbance was defined to be the sum of the absorbances at 2950, 2958 and 2966 cm^{-1} (6,7). The total absorbance was the sum of the absorbances for all of the wavenumbers in Table 1. The resulting calibration plot is shown in Figure 3. Copolymer concentrations for LLDPE and ULDPE resins were assumed to be proportional to the concentration of methyl groups. This assumption was made because in most cases molecular weight corrections for chain ends were small (see discussion).

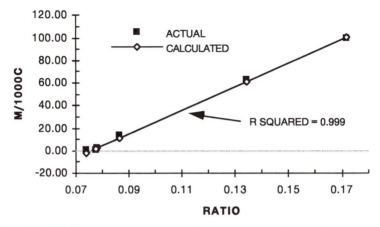

Fig. 3. GPC/FT-IR calibration plot of methyls per thousand carbons versus absorbance ratio (the absorbance ratio is the ratio of the methyl absorbance divided by the total absorbance). The plot contains actual and calculated values for the standards used in the calibration.

Quantitative Determination of Polar Comonomer Concentration. The polar comonomers in ethylene acrylic acid and ethylene carbon monoxide copolymers each possess a carbonyl functionality. Therefore, the carbonyl absorbance was used to determine the comonomer concentrations. The comonomer concentration was reported on a relative basis as the ratio of carbonyl absorbance divided by the total absorbance. The carbonyl absorbance was the sum of the absorbances at 1706 and 1714 cm^{-1}. The total absorbance was the sum of the absorbances for all of the wavenumbers in Table 1. Since spectra were taken at 40 equally spaced times during the GPC analysis; we were able to determine the polar comonomer concentration ratio at each of these times.

Results and Discussion

Figures 4, 5 and 6 illustrate the type of data which was obtained via GPC/FT-IR. Figures 4 and 5 are graphs of the molecular weight distribution (weight% versus Log molecular weight) and the methyls/1000C carbons distribution (methyls/1000C carbons versus log molecular weight) for two ULDPE resins. All of the molecular weight determinations were based on a single elution volume to molecular weight relationship (intended for linear polyethylene systems) as described previously. No attempt was made to correct for structural deviations from linear polyethylene. Therefore, the molecular weight data reported herein are meant for relative comparison between similar copolymers only. The reported branching data is uncorrected for molecular weight. Attempts to correct for chain ends using the molecular weight resulted in over correction (negative methyls/1000C) at low molecular weights. Alternative corrections using only one methyl termination per chain (assuming vinyls resulted from elimination reactions) gave more reasonable results. However, since the correction was only important for a very small percentage of the molecular weight distribution (< 10%) and since other effects, such as column resolution could not be ruled out, no correction was included. The methyls/1000C was plotted over a molecular weight range which corresponds to 97% of the polymer. The methyls/1000C data for points corresponding to molecular weights outside the range shown was subject to significant error because of the extremely low sample concentrations. The figures 4 and 5 indicate that the two resins have similar molecular weight distributions but very different methyls/1000C distributions.

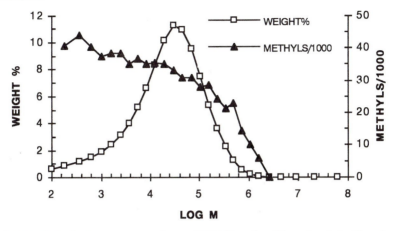

Fig. 4. GPC/FT-IR chromatogram for an ULDPE resin. Plot of weight % and methyls /1000C versus log molecular weight.

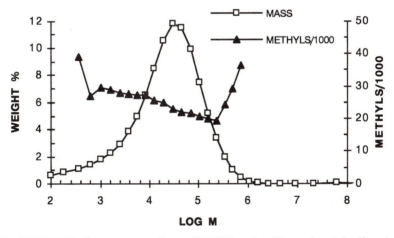

Fig. 5. GPC/FT-IR chromatogram for an ULDPE resin. Plots of weight % and methyls /1000C carbons versus log molecular weight.

Figure 6 is a graph of the molecular weight distribution and the acrylic acid comonomer distribution for an EAA resin. The acrylic acid distribution is represented as the ratio of the carbonyl absorbance divided by the total absorbance. The CO absorbance/total absorbance was plotted over a molecular weight range which corresponds to 97% of the polymer. Just as for the methyls/1000C data in figures 4 and 5, the carbonyl ratio data for points outside this molecular weights range were not shown because they were subject to significant error due to the extremely low sample concentrations.

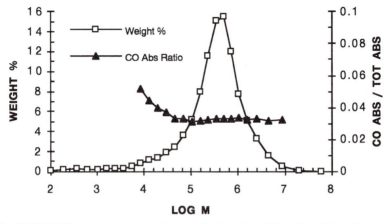

Fig. 6. GPC/FT-IR chromatogram for an EAA resin. Plot of weight % versus log molecular weight and of the carbonyl absorbance ratio (the ratio of the carbonyl absorbance divided by the total absorbance) versus log molecular weight.

Conclusions

The results demonstrate that GPC/FT-IR can be used to characterize ethylene based copolymers. GPC/FT-IR provides three key advantages:

1. One is able to simultaneously characterize molecular weight, molecular weight distribution, and comonomer concentrations (as a function of molecular weight).
2. The technique is much faster and simpler than alternative techniques; for example, preparative GPC fractionation and sample collection followed by infrared analysis and solid state IR versus solution IR.
3. The method is adaptable to many copolymer systems (or any other IR active functional groups) .

The data obtained from the analyses are solid evidence for claims concerning polymer structure, especially where molecular weight is a determining factor.

Literature Cited

1. Moore, J. C. *J. Polym. Sci.*, Part A2, **1964**, 835.
2. Brehon, F. *Spectra 2000 (Deux Mille)*, **1990**,*145*, 48.
3. Housaki, T.; Satoh, K.; Nishikida, K.; Morimoto, M. *Makromol. Chem. Rapid Commun.*, **1988**, *9*, 525.
4. Nishikida, K.; Housaki, T.; Morimoto, M.; Kinoshita, T. *J. Chromatogr.*, **1990**, *517*, 209.
5. Williams, T.; Ward, I. *J. Polym. Sci., Polym. Letters*, **1968**,*Vol. 6*, 621.
6. Bellamy, L. J. *The Infrared Spectra of Complex Molecules*, Wiley, New York 1975.
7. Bellamy, L. J. *Advances in Infrared Group Frequencies* , Barnes and Nobel, New York 1968.

RECEIVED August 27, 1992

Chapter 19

Molecular-Weight Determination of Poly(4-methyl-1-pentene)

Arja Lehtinen[1] and Hannele Jakosuo-Jansson[2,3]

[1]Polyolefins R&D, Neste Chemicals, P.O. Box 320, SF−06101 Porvoo, Finland
[2]Department of Chemical Engineering, Technical University of Helsinki, Kemistintie 1, SF−02150 Espoo, Finland

Molecular weight studies of poly-4-methyl-1-pentene were carried out by high temperature SEC and viscosity measurements. For SEC analyses, the samples were dissolved in a TCB/Decalin mixture by microwave heating and run at 135 °C with TCB as eluent. The conditions chosen were based on studies of solubility of PMP in different solvents, stabilization of the solutions with different anti-oxidants and dissolution experiments in an air oven and a microwave oven. For the system calibration the samples were analyzed both by SEC and SEC/on-line viscometer combination. Results for a set of isotactic poly-4-methyl-1-pentene samples with different melt flow rates are presented. Very high molecular weight poly-4-methyl-1-pentenes were found to require a very careful procedure for measurement of reliable molecular weight averages.

Isotactic poly-4-methyl-1-pentene (PMP) is a semicrystalline polyolefin with a bulky side group. It is the most heat resistant of the polyolefins produced today and may be used, in the absence of oxygen, at temperatures up to 180 - 200 °C. Other attractive properties are its low density, resistance to chemicals, microwavability, transparency and recyclability. The application areas of PMP include food packaging, medical equipment, auto parts, release paper, wire and cable coating, laboratory and micro-oven ware (*1*).

Determination of molecular weight distribution and molecular weight averages is an essential part of polymer characterization. Today, size exclusion chromatography (SEC) is the most convenient and most informative method for these measurements. Numerous papers about the high temperature SEC analyses of polyethylene and polypropylene have been published (*2-6*). As compared with these

[3]Current address: Technology Centre, Neste OY, P.O. Box 310, SF−06101 Porvoo, Finland

0097−6156/93/0521−0277$06.00/0

two other polyolefins, poly-4-methyl-1-pentene is yet a less commonly known polymeric material. The studies on it have mainly concentrated on its crystallization behaviour and morphology. Only a few references about its molecular weight analysis are available in the literature (7-10). The poor solubility of PMP and its sensitivity to oxidative degradation have made these determinations difficult.

The work reported here was undertaken to find appropriate experimental conditions for the SEC analyses of poly-4-methyl-1-pentene. Therefore, in addition to high temperature SEC runs, solubility of PMP in different solvents, stabilization of the solutions with different antioxidants and dissolution of samples with hot air and microwave heating were also studied.

Experimental Details

Materials. Three of the samples used in this study (A,B & D) were homopolymers. Their melting temperature was 236 °C. The fourth sample was a copolymer containing 1.9 wt-% of 1-decene (T_m = 231 °C). All samples were in pellet form.

Instruments. The SEC measurements were carried out using four Waters 150C ALC/GPC instruments. A Viscotek Model 100 differential viscometer was coupled to one of the instruments (in parallel with the RI detector). The eluting solvent was vacuum distilled 1,2,4-trichlorobenzene (TCB). The operating temperature was 135 °C, flow rate 1 ml/min and injection volume 250 - 500 μl. Antioxidants were added only to sample solutions (75 - 125 mg in 300 ml of TCB). The column sets used in this study consisted either of three TSK-Gel columns or four Shodex columns (Table I). Depending on the SEC instrument used, data acquisition and analysis were performed either with Millipore/Waters Maxima 820, Polymer Laboratories or in-house modified Viscotek SEC software. The columns were calibrated with seventeen NMWD polystyrene standards obtained from Polymer Laboratories (their M_p ranging from 750 to 9.8 x 10^6 g/mol). The PS calibrations obtained were approximated by third order polynomial fits and checked with NBS 706 BMWD PS standard (the universal calibration of the SEC/DV system was also verified with BMWD HDPE standards).

Sample Preparation. The preliminary dissolution experiments of the samples were made in small erlenmeyer flasks using a heatable magnetic stirrer and visual observation. The sample   concentration was about 0.7 mg/ml. For the SEC analysis, the solutions were prepared directly in the 4 ml glass vials of the injector. Depending on the melt flow rate of the sample, 1.5 - 2.1 mg of polymer was weighed in the vial and then 3 ml of solvent was added. The vials were closed with a Teflon septum and screw cap, then placed either in an air oven set at 150 °C or in the Teflon vessels of the microwave oven (CEM DMS-81). The solutions were analyzed without any filtering.

Viscosity Measurements. Off-line intrinsic viscosities were measured in decahydronaphthalene (Decalin) at 135 °C with a Viscotek Model 100 differential viscometer. The system was calibrated with five polyethylene standards with known viscosity values in Decalin. The PMP samples were dissolved at 150 °C for 2 - 3 hrs using a heatable magnetic stirrer and then thermostated at 135 °C for half an hour. Sample concentration was 20 - 40 mg or less polymer in 100 ml of Decalin and several solutions were measured from each sample. All solutions were stabilized with Irganox B215. The melt flow rates of the PMP samples (MFR₂) were measured with a Ceast CE-UM-118 melt indexer at 260 °C.

Table I. Column parameters

Instrument	Column set	Plates/m	NBS 706 $\overline{M}_w$x10^{-3}	$\overline{M}_w/\overline{M}_n$
SEC I	TSK-Gel 2 x GMHXL-HT + 7000HXL	19 800	300	2.2
SEC II	TSK-Gel 3 x GMHXL-HT	14 600	260	2.1
SEC III	TSK-Gel 2 x GMHXL-HT + 7000HXL	13 000	295	2.2
SEC/DV	Shodex A-807/S + 2 x AT-80M/S + A-803/S	31 600	299	2.1

Results and Discussion

The molecular weight determinations of polymers usually require the polymer to be dissolved in a suitable solvent. Therefore, some preliminary experiments on the solubility of poly-4-methyl-1-pentene in solvents suitable for SEC and viscosity measurements were carried out. Since the solubility of the polymers depends to some extent on molecular weight, the highest molecular weight materials being the most difficult to dissolve, samples C and D having the lowest melt flow rates were chosen for these studies. In a recent study (*11*), PMP was dissolved in solvents such as chloroform, carbon tetrachloride, cyclopentane, cyclohexane, p-xylene and cis-Decalin for film casting. The dissolution was achieved within 8 hrs. In our experiments, the dissolution time was 1 hr and the solvents used are listed in Table II. Decalin, TCB and methylcyclohexane were found to be the best solvents for our PMP samples, the density of PMP (about 0.835 g/cc) being quite near that of Decalin and methylcyclohexane. The samples were insoluble in cyclohexane, toluene and xylene under the conditions used.

In the two SEC methods found in the literature, poly-4-methyl-1-pentene was analyzed either in TCB at 145 °C (*7*) or in methylcyclohexane at 80 °C (*8*). Since TCB is the most commonly used solvent in the SEC analyses of other polyolefins, it was preferred. A very high operating temperature in SEC shortens the column lifetime and increases the possibility of polymer degradation (*2,12*), therefore, an

Table II. Results of the dissolution experiments

Solvent	Boiling temp. °C	Density g/cc	Solubility of PMP[*)]
Cyclohexane	81	0.78	-
Methylcyclohexane	98-101	0.77	+ (80 °C/40 min)
Toluene	111	0.87	-
Xylene	139-144	0.87	-
Decalin	187-196	0.88	+ (140 °C/0.5 hr)
o-Dichlorobenzene (ODCB)	180	1.31	+ (170 °C/0.5 hr) precip. at 100 °C
Trichlorobenzene (TCB)	213	1.57	+ (150 °C/0.5 hr) precip. at 135 °C
1-Chloronaphthalene	259	1.19	+ (180 °C/0.5 hr) precip. at 150 °C

[*)] - = insoluble, + = soluble (temperature/dissolution time)

analysis temperature lower than 145 °C was desirable for PMP. The poly-4-methyl-1-pentene macromolecule contains in monomer unit two tertiary carbon atoms, one in the main chain and the other in the side group, which makes it very sensitive to thermo-oxidative degradation (7,13).

In the preliminary dissolution experiments, PMP started to precipitate from TCB solution when the temperature was slowly lowered below 140 °C. How rapidly the solution became turbid, seemed to depend on the nature and amount of the antioxidant used to stabilize it. In our laboratory, the high temperature SEC instruments are normally operated at 135 °C. At this temperature the precipitation was slowest when the antioxidant was Ionol (= BHT). This antioxidant has a low molecular weight and a low melting temperature as shown in Table III. Since antioxidant was added only to sample solutions, peaks of the antioxidants having higher molecular weight (e.g. Irganox 1010 and B225) also caused problems in the molecular weight calculations by overlapping with the low-molecular weight tail of some PMP samples. Having a column with a small pore size in each column set might have solved this problem. In high temperature SEC the life-time of this kind of column is short, and consequently only the column set of SEC/DV contained such a column.

To minimize potential degradation of the polymer, the high temperature exposure time has to be kept as short as possible in the SEC analyses of PMP (7). Polypropylene is also sensitive to oxidative degradation and difficult to dissolve.

Table III.   Antioxidants used in the SEC trial runs (melting temperatures determined by DSC)

Antioxidant	MW (g/mol)	T_m (°C)
N-phenyl-2-naphthylamine	219	108.4
Ionol	220	70.6
Santonox R	358	162.9
Irganox 1010	1178	116.2
Irganox B225*)		116/179

*) an 1:1 mixture of Irganox 1010 and Irgafos 168.

However, it can be analyzed by SEC at temperatures as low as 60 - 80 °C using cyclohexane as eluent and dissolving the samples in a mixture of Decalin and cyclohexane (*14,15*). In this study, it was found that PMP could be analyzed at 135 °C using TCB as eluent when the samples were dissolved in a 9:1 mixture of TCB and Decalin. The dissolution temperature was 150 °C and the dissolution time in an air oven was 1 hr. When only TCB was used, 3 hrs were needed for the dissolution of the sample. The PMP-TCB-Decalin solutions could be kept at 135 °C overnight without appearance of any turbidity.

The dissolution time could further be reduced to about 10 minutes when the samples were dissolved by microwave heating in the 4 ml glass vials of the injector. The exact times needed for dissolution varied somewhat depending on the melt flow rate of the sample and the antioxidant used. The microwave heating raised the temperature in the solutions up to around 170 °C, and therefore after dissolution the sample solutions were allowed to stand in the SEC instrument for about an hour before the first PMP injection was made.

According to the trial runs in Figure 1, the small amount of Decalin added to the PMP-TCB solution did not accelerate the degradation of the polymer. In these trial runs the antioxidant concentration was about 0.4 mg/ml and each injection was made from a different solution. The critical time for the stability of the PMP solutions seemed to be around 6 hrs. The SEC runs were planned so that the total run time was below this time. The first sample in every run was a LLDPE reference sample and usually only one injection from each solution was made.

Even if the dissolution time of PMP was very short when microwave heating was used, no problems with system pressure, column blocking or dramatic changes in the plate count of the columns were encountered in the SEC analyses. Since the absence of visual turbidity was no guarantee that the PMP samples had dissolved in an aggregate free state, intrinsic viscosity versus molecular weight from the SEC/DV data was plotted for each sample. As the example in Figure 2 shows these log-log Mark-Houwink's plots were reasonably linear into the high molecular weight region indicating true solution of the samples.

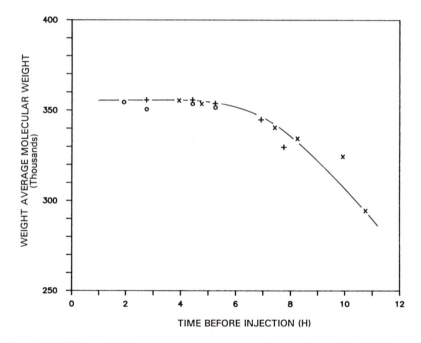

Figure 1. Effect of high temperature exposure time and Decalin addition on the molecular weight of PMP sample A (SEC II, $\overline{M}_w$ in PS equivalents; + - TCB/Ionol, o - TCB/Decalin/Ionol, x - TCB/Decalin/Irganox 1010).

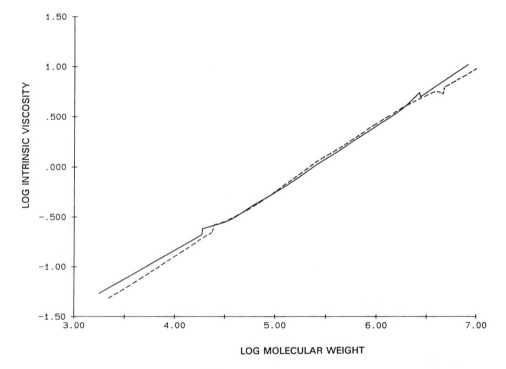

Figure 2. Log [η] versus log $\overline{M}_w$ plots for samples A (——) and B (---) (from SEC/DV data).

When the molecular weight averages determined by SEC were compared, no statistical difference could be observed between the data obtained by the two different dissolution techniques (air oven vs. microwave oven). Similar results have also been obtained for polyethylene and polypropylene (16). Taking into account the small sample amount used and the normal pellet to pellet inhomogeneity of the samples, the repeatability of the analyses was good ($\bar{M}_n$ and $\bar{M}_w$). In the example in Table IV, the sample was dissolved by microwave heating in the TCB-Decalin mixture. All injections were made from different solutions.

Table IV.   Repeatability of the SEC runs of PMP
(sample B, SEC/DV)

$\bar{M}_n \times 10^{-3}$	$\bar{M}_w \times 10^{-3}$	$\bar{M}_z \times 10^{-3}$	$\bar{M}_w/\bar{M}_n$
91.0	397	1 430	4.4
91.2	393	1 110	4.3
87.2	393	1 250	4.5
85.6	391	1 200	4.5
89.3	407	1 280	4.6
88.9	396	1 250	4.6

A SEC system without any molecular weight detector has to be calibrated for molecular weight determinations. In the case of poly-4-methyl-1-pentene this is problematic since no commercial PMP standards are available. The only real SEC study of PMP published (7) gave the results in polystyrene equivalents. The calibration in another paper (8) is unknown. All our data handling was also first done using a NMWD polystyrene calibration. For the samples A, B and C, the SEC analyses were highly reproducible. The three SEC instruments with different data handling systems and columns gave very consistent molecular weight values. In the SEC instrument II which only had three mixed bed columns, the exclusion limit of the columns was too low for the analysis of higher molecular weight PMP. The large deviations in the molecular weights obtained for the sample D may also be partly due to different exclusion limits of the columns. SEC I and SEC III had similar column sets, but the PMP samples were analyzed at different points of their lifetime (the plate counts in Table I, analysis of PMP after 250 vs. 650 injections of other polyolefins). The exclusion limit of the column set normally decreases when the plate count starts to decrease.

To get some idea of the real molecular weight averages of poly-4-methyl-1-pentene, all samples were also analyzed with a SEC/on-line viscometer combination (SEC/DV). In this combination system, the intrinsic viscosity is determined as a function of molecular weight and the molecular weight averages are calculated using universal calibration. The Mark-Houwink equation relates the intrinsic viscosity of a polymer to its viscosity average molecular weight through the empirical constants K and a. This intrinsic viscosity-molecular weight relationship can be determined by SEC/DV as shown in Figure 2. Based on the SEC/DV data of PMP samples A, B and C, and NMWD PS standards used in the calibration, the following Mark-Houwink equations were obtained for poly-4-methyl-1-pentene and polystyrene in TCB at 135 °C:

PMP: $[\eta] = 3.1 \times 10^{-4} \ (M)^{0.66}$

PS: $[\eta] = 2.31 \times 10^{-4} \ (M)^{0.66}$

The SEC chromatograms were recalculated using universal calibration and these parameters. The resulting molecular weight data together with SEC/DV data are presented in Table V and the molecular weight distributions curves in Figures 3 and 4. The universal calibrations of SEC I and SEC/DV gave surprisingly similar molecular weight averages for samples A, B and C, even if the plate counts and separation ranges of the columns were different (Table I, Figure 5). Thus the Mark-Houwink's constants obtained by SEC/DV seemed to be reasonable enough for practical use. The columns in SEC/DV had a high plate count and the data handling software included a peak broadening correction. This may explain the narrower molecular weight distributions. In SEC II and SEC III, the column plate count had already decreased somewhat during the earlier runs and the constants did not fit as well. The Mark-Houwink's constants derived from SEC depend on the calibration range of the samples and on the column plate count (2,17).

Table V. Comparison of molecular weight averages obtained by SEC/DV and by SEC using universal calibration and Mark-Houwink's constants obtained by SEC/DV

Sample		A	B	C	D
SEC I	$\bar{M}_n \times 10^{-3}$	71 000	75 000	85 000	122 000
	$\bar{M}_w \times 10^{-3}$	310 000	401 000	556 000	950 000
	$\bar{M}_w/\bar{M}_n$	4.4	5.3	6.5	7.8
SEC II	$\bar{M}_n \times 10^{-3}$	74 000			
	$\bar{M}_w \times 10^{-3}$	300 000			
	$\bar{M}_w/\bar{M}_n$	4.1			
SEC III	$\bar{M}_n \times 10^{-3}$	64 000	72 000	90 000	135 000
	$\bar{M}_w \times 10^{-3}$	295 000	355 000	562 000	654 000
	$\bar{M}_w/\bar{M}_n$	4.6	4.9	6.3	4.8
SEC/DV	$\bar{M}_n \times 10^{-3}$	88 000	89 000	100 000	71 000
	$\bar{M}_w \times 10^{-3}$	320 000	396 000	565 000	655 000
	$\bar{M}_w/\bar{M}_n$	3.6	4.6	5.7	9.2

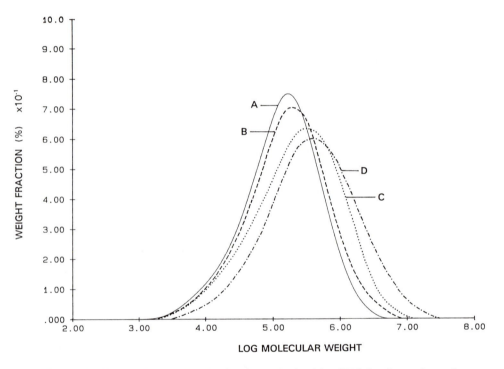

Figure 3. Molecular weight distributions obtained by SEC I using universal calibration.

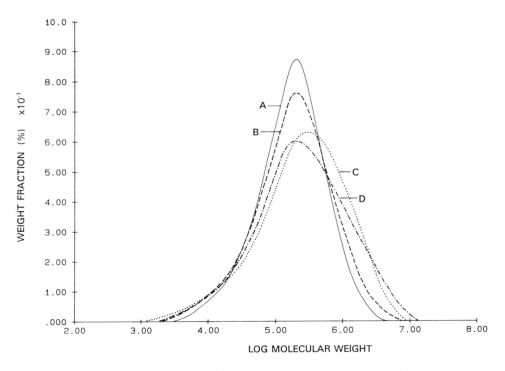

Figure 4. Molecular weight distributions obtained by SEC/DV.

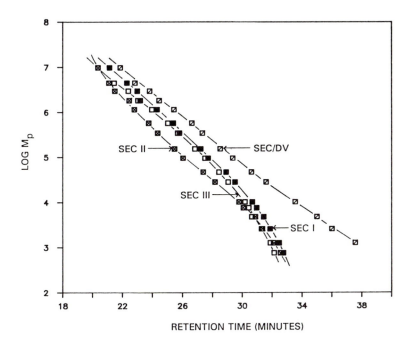

Figure 5. Polystyrene calibrations of the instruments.

Table VI. Viscosity average molecular weights obtained by SEC/DV and off-line viscosity measurements

Sample	MFR$_2$	SEC/DV		Off-line visc. measurements[*]		
		$[\eta]_{TCB}$	$\bar{M}_v \times 10^{-3}$	$[\eta]_{DEC}$	$\bar{M}_v \times 10^{-3}$	$\bar{M}_v \times 10^{-3}$
A	19.8	0.98	264	1.48	62	149
B	11.8	1.11	317	1.63	70	170
C	2.45	1.58	449	2.19	101	252
D	0.23	1.74	437	3.28	166	431

[*] $\bar{M}_v$ calculated using following Mark-Houwink's constants (9,10):
$K = 1.94 \times 10^{-4}$ & a = 0.81 and $K = 1.95 \times 10^{-4}$ & a = 0.75.

Since the viscosity-molecular weight relationship of poly-4-methyl-1-pentene has been established in Decalin at 130 and 135 °C (9,10), the intrinsic viscosities of the samples were measured in this solvent at 135 °C. Table VI shows the on-line and off-line viscosities measured for the PMP samples studied and the viscosity average molecular weights ($\bar{M}_v$) calculated using them. For the samples A, B and C, the intrinsic viscosity in Decalin was about 1.5 times that in TCB. The off-line viscosity measurements in Decalin gave much lower $\bar{M}_v$ values than the SEC and SEC/DV measurements in TCB. The reliability of the Mark-Houwink's constants of PMP in Decalin is unknown. These constants were determined using polydisperse samples and their M_n values.

To obtain good mechanical properties, the polymer molecules have to have a certain length. Because of the bulky side group, to obtain the same chain length as with polyethylene, the molecular weight of PMP has to be three times that of polyethylene. Compared with normal HDPE the molecular weights of poly-4-methyl-1-pentenes may be very high. One extreme example was the high molecular weight PMP sample D. Because of the scattering of its SEC results and certain inconsistencies in the viscometric and melt flow rate data, the off-line viscosities and melt flow rates of all samples were remeasured paying careful attention to the experimental parameters (e.g. dissolution conditions and concentrations in the viscosity measurements). A higher intrinsic viscosity and lower melt flow rate than the original values were obtained for sample D in these new measurements (18). The new viscosity and melt flow rate values correlated well with each other as shown in Figure 6. This indicates that the HMW PMP sample D, which was very difficult to melt or dissolve completely, might have degraded partially during the original measurements.

A comparison between intrinsic viscosities measured in Decalin and weight average molecular weights determined by SEC I and SEC/DV (Figure 7) shows that the weight average molecular weight of sample D determined by SEC/DV was probably too low. High molecular weight polymers are very sensitive to shear

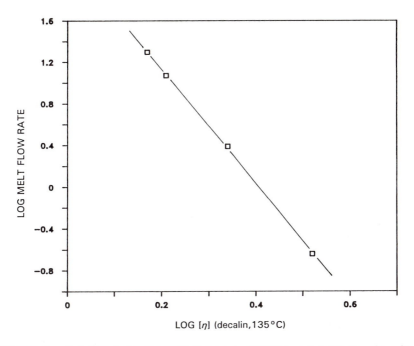

Figure 6. Relation between melt flow rate (MFR$_2$) and intrinsic viscosity measured in Decalin.

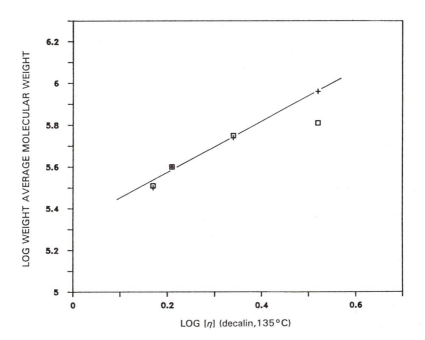

Figure 7. Log M$_w$ vs. log [η]$_{DEC}$ (+ SEC I, □ SEC/DV).

degradation (19). For the sample D, the concentration used was low in all SEC analyses (about 0.5 mg/ml) and no detectable influence of possible pressure increase was seen in the shapes or places of the peaks of the on-line chromatograms. However, the flow rate used (1 ml/min) was too high for this type of polymer. In addition, the SEC/DV system contained more capillaries than normal SEC and the Shodex columns used had slightly smaller particle size than the TSK-Gel columns, so increased shear degradation was possible. Partial degradation of the polymer would also explain the low intrinsic viscosity and the broader molecular weight distribution obtained for this sample by SEC/DV.

Conclusions

A high temperature SEC method for the determination of molecular weight distribution of poly-4-methyl-1-pentene has been developed. In this method, the samples are dissolved in a 9:1 mixture of TCB and Decalin using microwave heating and run by SEC at 135 °C with TCB as eluent. Even a lower operating temperature might be possible, but it was not studied. The antioxidant chosen to stabilize the polymer solutions was Ionol (BHT).

Analysis of very high molecular poly-4-methyl-1-pentene was found to be complicated. More experimental work is needed in this area. For the lower molecular weight samples the molecular weight determinations were repeatable and reproducible. Partial degradation of the samples was detected when the solutions were kept for long times at high temperature (over 6 hrs).

Compared with the universal calibration, polystyrene calibration gave too high molecular weight averages, especially for the higher molecular weight PMP samples. The viscosity average molecular weights measured in TCB and Decalin differed from each other very much. Viscosity average molecular weight is influenced by the solvent and temperature used in its determination. However, to eliminate possible inaccuracies in the Mark-Houwink's constants of PMP in Decalin and TCB, these constants should be determined using NMWD fractions and $\overline{M}_w$ measured by light scattering.

Acknowledgement

The authors are grateful to Mrs Raija Vainikka for helping with experimental details and for the SEC/DV analyses.

Literature Cited

1. Meinander, G. *Kemia-Kemi* **1989**, *16*, 911.
2. Lehtinen, A.; Vainikka, R. *Proc. Int. GPC Symp. '89*, Waters Chromatography Division, Millipore Corp., Milford, MA, 1989, pp. 612-623.
3. Grinshpun, V. *Ph. D. Thesis*, University of Waterloo, Waterloo, Ontario 1986.
4. Haddon, M. R; Hay, J. N. in *Size Exclusion Chromatography*, Hunt, B. J.; Holding, R. S., Eds.; Blackie and Son Ltd, London, 1989, pp. 56-99.
5. Lew, R,; Suwanda, D.; Balke, S. T. *J. Appl. Polym. Sci.* **1988**, *35*, 1049.

6. Lew, R.; Suwanda, D.; Balke, S. T. *J. Appl. Polym. Sci.* **1988**, *35*, 1065.
7. Fuller, E. N.; Edwards, G. Jr. *Proc. Int. GPC Symp.* *'89*, Waters Chromatography Division, Millipore Corp., Milford, MA, 1989, pp. 624-636.
8. Neuenschwander, P.; Pino, P. *Makromol. Chem.* **1980**, *181*, 737.
9. Tani, S.; Hamada, F.; Nakajima, A. *Polym. J.* **1973**, *5*, 86.
10. Hoffman, A. S.; Fries, B. A.; Condit, P. C. *J. Polym. Sci.* **1963**, *C4*, 109.
11. Mohr, J. M.; Paul, D. R. *Polymer* **1991**, *32*, 1236.
12. Wild, L.; West, B. *Proc. Int. GPC Symp.* *'89*, Waters Chromatography Division, Millipore Corp., Milford, MA, 1989, pp. 357-372.
13. Yasina, L. L.; Pudov, V. S. *Polym. Sci. U. S. S. R.* **1982**, *24*, 541.
14. Ying, Q.; Xie, P.; Ye, M. *Makromol. Chem., Rapid Commun.* **1985**, *6*, 105.
15. Ibhadon, A. O. *J. Appl. Polym. Sci.* **1991**, *42*, 1887.
16. Gilman, L. B.; Dark, W. A. *Proc. Int. GPC Symp.* *'89*, Waters Chromatography Division, Millipore Corp., Milford, MA, 1989, pp. 496-507.
17. Dobbin, C. J. B.; Rudin, A.; Tchir, M. F. *J. Appl. Polym. Sci.* **1980**, *25*, 2985.
18. Lehtinen, A.; Jakosuo, H. *Polym. Mater. Sci. Eng.* **1991**, *65*, 93.
19. Barth, H. G.; Carlin, F. J., Jr. *J. Liq. Chromatogr.* **1984**, *7*, 1717.

RECEIVED July 6, 1992

Chapter 20

Size-Exclusion Chromatography of Cationic, Nonionic, and Anionic Copolymers of Vinylpyrrolidone

Chi-san Wu, James F. Curry, and Lawrence Senak

International Specialty Products, 1361 Alps Road, Wayne, NJ 07470

A review of size exclusion chromatography (SEC) data from our laboratory for cationic, nonionic, and anionic copolymers of vinylpyrrolidone is presented. Cationic copolymers, quaternized poly(vinylpyrrolidone-co-dimethylaminoethylmethacrylate), PVPDMAEMA, were investigated using a 0.1M TRIS buffer, pH 7, 0.5M LiNO$_3$ mobile phase with Waters Ultrahydrogel columns. With these conditions, weight-average molecular weights of PVPDMAEMA determined by both SEC/LALLS and SEC with universal calibration were in good agreement. SEC of nonionic copolymers, poly(vinylpyrrolidone-co-vinylacetate), PVPVA, and poly(vinylpyrrolidone-co-dimethylaminoethylmethacrylate-co-vinylcaprolactam), PVPDMAEMAVC, was studied using an aqueous mobile phase with a four-column set and an organic mobile phase with various two-column sets. SEC results with these conditions were evaluated qualitatively in terms of resolution between polymer and solvent peaks. Vinylpyrrolidone compositions of PVPVA ranging from 30 to 70 mole% were studied. SEC of anionic copolymers, poly(vinylpyrrolidone-co-acrylic acid), PVPAA, at pH 9, was studied using a 0.1M TRIS buffer, pH 9, 0.2M LiNO$_3$ with Waters Ultrahydrogel columns. PVPAA with vinylpyrrolidone compositions ranging from 25 to 90 mole% were studied.

Various copolymers of vinylpyrrolidone, both neutral and ionic in character, have been studied using SEC. The purpose of this paper is to review the SEC results on these water soluble copolymers. Workers had previously found a satisfactory mobile phase for SEC of nonionic poly(vinylpyrrolidone), PVP, homopolymers (1). This mobile phase, a 1:1 water/methanol solution containing 0.1M LiNO$_3$ was found to cause elution of PVP based on hydrodynamic volume using Toyo Soda TSK-PW columns (similar in packing material to Waters Ultrahydrogel

0097–6156/93/0521–0292$06.00/0

columns) (*2*). A log-linear molecular weight separation of PVP spanning over three decades (10^3 to 10^6 g/mol) was obtained. Since the water/methanol mobile phase was useful for the SEC of PVP, this mobile phase was used to study two nonionic copolymers of PVP; poly(vinylpyrrolidone-co-vinylacetate), PVPVA, and poly(vinylpyrrolidone-co-dimethylaminoethylmethacrylate-co-vinylcaprolactam), PVPDMAEMAVC. It was of general interest to also explore SEC of PVPVA and PVPDMAEMAVC using a nonaqueous solvent, and dimethylformamide (DMF) was the solvent of choice because the copolymers were soluble in it and because of the polarity of DMF relative to water/methanol.   Cationic copolymers of vinylpyrrolidone, quaternized poly(vinylpyrrolidone-co-dimethylamino-ethylmethacrylate), PVPDMAEMA, have also been studied using SEC. The quaternized amino groups on these copolymers are responsible for the cationic charge at pH 7. Weight-average molecular weights of PVPDMAEMA, determined by both universal calibration and SEC with low angle laser light scattering (SEC/LALLS) were found to be in good agreement (*3*). The SEC conditions using a 0.1M TRIS, pH 7 buffer, 0.5M $LiNO_3$ mobile phase on Waters Ultrahydrogel columns were found to cause elution of PVPDMAEMA based on hydrodynamic volume.

SEC results for compositions of anionic poly(vinylpyrrolidone-co-acrylic acid), PVPAA, using four Waters Ultrahydrogel columns with 0.1M TRIS buffer, pH 9, 0.1M $LiNO_3$ have also been reported (*4*). The mobile phase and columns were chosen in part becaue they were previously found, when combined, to yield universal calibration for an anionic copolymer, poly(methylvinylether-co-maleic anhydride) (*5*).

Experimental

PVPDMAEMA, PVPVA, PVPDMAEMAVC, and PVPAA, all synthesized by free radical polymerization, were produced by International Specialty Products (formerly GAF Chemicals Corp.). Single lots of each grade of the four copolymers were used. Three grades of PVPDMAEMA samples were obtained: a low-molecular-weight grade polymer to be referred to as 734, a high-molecular-weight grade polymer to be referred to as 755, and a neutralized high-molecular-weight grade polymer to be referred to as 755N. The composition and synthesis of these copolymers have been previously reported (*6-7*). PVPVA samples were obtained in the following form: a 50% solution in ethanol, E series; a 50% solution in isopropanol, I series; a 50% solution in water, W series; and a free flowing powder, S series. Vinylpyrrolidone compositions of PVPVA were expressed by the first digit of the sample grade label. For example, PVPVA sample grade I535, is 50 mole% vinylpyrrolidone, in isopropanol, and PVPVA sample grade E735 is 70 mole% vinylpyrrolidone, in ethanol. Vinylpyrrolidone compositions of PVPVA ranging from 30 to 70 mole% were studied. PVPAA samples were obtained as a free flowing powder. PVPAA sample grades studied, 1005, 1004, 1001, and 1030, had 25, 50, 75, and 90 mole% vinylpyrrolidone, respectively. PVPDMAEMAVC samples were obtained as a 37% solution in ethanol. The composition and synthesis of this polymer has been discussed in an earlier patent (*8*).

PVPDMAEMA samples were prepared as 0.25% (w/v) solutions in tris-hydroxymethylaminoethane (TRIS) with 0.5M LiNO$_3$, and adjusted to pH 7 using HNO$_3$. This TRIS solution, to be referred to as pH 7 buffer, was used as the mobile phase for the SEC study on PVPDMAEMA. SEC conditions for PVPVA and PVPDMAEMAVC were investigated using either an aqueous or an organic mobile phase. A 1:1 water/methanol (v/v), 0.1M LiNO$_3$ solution, to be referred to as aqueous mobile phase, or a DMF, 0.1M LiNO$_3$ solution, to be referred to as organic mobile phase, was used. Sample solutions of the three copolymers noted, with concentrations of 0.1% (w/v), were placed on a slowly rotating wheel for 1/2 day to achieve dissolution. For PVPAA, 1% (w/v) solutions in 0.25M NaOH were prepared since the polymer dissolves readily in this alkaline solution. The alkaline PVPAA solutions were then diluted 1:9 (v/v) using 0.1M TRIS, 0.2M LiNO$_3$ solution, previously adjusted to pH 9 with HNO$_3$. This TRIS solution will be referred to as pH 9 buffer. A column set consisting of four Waters Ultrahydrogel columns, 120, 500, 1000, and 2000, was used for each of the four types of copolymers and the corresponding aqueous mobile phases specified. This column set will be referred to as U4. Three different two-column sets were used for PVPVA and PVPDMAEMAVC studied using organic mobile phase. These three column sets, Shodex KD80M plus Ultrahydrogel 120, Shodex KD80M plus PLgel 100, and PLgel 10^4 plus PLgel 500, will be referred to as SU2, SP2, and PP2, respectively. The procedure of adding a Waters Ultrahydrogel column in series after a PLgel column was recently reported (9).

Intrinsic viscosity, [η], for the copolymers was determined at 25°C using a 0.63mm Ubbelohde viscometer with the solvents employed as mobile phases for the SEC experiments. For PVPDMAEMA, PVPVA, PVPDMAEMAVC, and PVPAA studied by SEC in aqueous solution, concentrations between 1.0 and 0.2 g/dl were used to determine [η] by extrapolation of the reduced viscosity to zero concentration. In order to determine [η] for PVPVA in organic solution, a single concentration was used. [η] was calculated using

$$[\eta] = \frac{(2(\eta_{sp} - \ln\eta_r))^{1/2}}{c} \tag{1}$$

where η_{sp} is specific viscosity, $\ln\eta_r$ is inherent viscosity, and c is concentration. Equation 1 was derived by subtracting (10)

$$\frac{\ln\eta_r}{c} = [\eta] + k''[\eta]^2 c \tag{2}$$

from (11)

$$\frac{\eta_{sp}}{c} = [\eta] + k'[\eta]^2 c \tag{3}$$

and substituting k' - k" = 1/2. Since a single concentration was used to determine [η] for PVPVA in organic solution, [η] for two of the PVPVA copolymer compositions was determined by extrapolation of reduced viscosity to zero concentration. [η] determined either by extrapolation or using Equation 1 was found to be in good agreement.

The SEC apparatus included a Waters model 715 or 710B WISP auto injector, a Waters model 590 pump, and a Waters model 410 differential refractometer. Injection volumes were typically 100 μl. Detector signals were collected on a DEC MINC-11 computer. Molecular weight information was obtained from the acquired data using GPC3 software from Chromatix, Inc. The Chromatix MOLWT3 software was also used to perform SEC/universal calibration calculations on the PVPDMAEMA samples. Polyethylene oxide (PEO) standards from Toyo Soda and Polymer Laboratories, with molecular weights between 860,000 and 7,100 g/mol, were used in all SEC experiments. These standards had polydispersities, ($\bar{M}_w/\bar{M}_n$), between 1.02 and 1.10.

Results and Discussion

Table I shows intrinsic viscosity data for PVPDMAEMA samples and PEO standards. Table II shows absolute molecular weights and molecular weight distributions of PVPDMAEMA samples determined from SEC/LALLS using the U4 column set with pH 7 buffer. Details of the light scattering experiments have previously been reported (*3*). Using the results shown in Tables I and II, the Mark-Houwink constants, K and a, were determined for PVPDMAEMA and PEO from plots of log(η) versus log(M). The Mark-Houwink constants are shown in Table III along with the coefficients of linear regression from the plots of log(η) versus log(M) for each polymer. Absolute molecular weights and molecular weight distributions for PVPDMAEMA were calculated by universal calibration using the PEO standards and these results are shown in Table IV.

Very good agreement was obtained for $\bar{M}_w$ determined by SEC/LALLS and SEC with universal calibration for the PVPDMAEMA samples. This suggests that the separation mechanism of both nonionic PEO and cationic PVPDMAEMA using pH 7 buffer is based on hydrodynamic volume. The agreement also supports an earlier finding (*2*) that the Mark-Houwink constants may be obtained from broad distribution polymers without fractionation for the purpose of calculating $\bar{M}_w$ by the universal calibration method.

For the SEC work on nonionic PVPVA and PVPDMAEMAVC, the dependence of retention volume with molecular weight for PEO standards eluted using the SU2 column set with the organic mobile phase is shown by curve a in Figure 1. A good linear correlation was obtained. Elution of the PEO standards on the SP2 and the PP2 column sets each showed a similar result. Curve b in Figure 1 shows the dependence of retention volume with molecular weight for PEO standards eluted using the U4 column set with the aqueous mobile phase. A good linear correlation was obtained. Figure 1 shows that better molecular weight resolution can be obtained using the aqueous system (four-column set) compared to the organic system (two-column sets). For all the copolymers studied in aqueous solution, Figure 2 represents a typical result obtained for fitted data of reduced viscosity versus concentration. This Figure shows a graph of η_{sp}/c versus concentration for PVPAA 1001 in pH 9 buffer.

Table I   Intrinsic Viscosity of PVPDMAEMA and PEO at 25°C in pH 7
buffer. PEO standards are listed by weight-average molecular
weight

Polymer	[η] (dl/g)
734	0.647
755	2.15
755N	2.22
PEO--	
860,000	5.353
270,000	2.418
160,000	1.568
85,000	0.970
45,000	0.592
21,000	0.374
12,600	0.265

Reprinted with permission from ref. 3. Copyright 1990 Marcel Dekker.

Table II   Absolute molecular weight and molecular weight distributions
of PVPDMAEMA from SEC/LALLS at 25°C in pH 7 buffer

PVPDMAEMA	$\bar{M}w$	$\bar{M}n$	$\bar{M}w/\bar{M}n$
734	3.00×10^5	1.15×10^5	2.61
755	1.63×10^6	7.04×10^5	2.32
755N	2.02×10^6	8.89×10^5	2.27

Reprinted with permission from ref. 3. Copyright 1990 Marcel Dekker.

Table III   Mark-Houwink constants and coefficient of linear regression
for PVPDMAEMA and PEO at 25°C in pH 7 buffer

Polymer	K	a	r
PVPDMAEMA	1.42×10^{-4}	0.67	0.9975
PEO	2.80×10^{-4}	0.72	0.9994

Reprinted with permission from ref. 3. Copyright 1990 Marcel Dekker.

Table IV   Absolute molecular weight and molecular weight distributions of
PVPDMAEMA from universal calibration at 25°C in pH 7 buffer

PVPDMAEMA	$\bar{M}w$	$\bar{M}n$	$\bar{M}w/\bar{M}n$
734	3.31×10^5	1.10×10^5	3.01
755	1.72×10^6	4.83×10^5	3.55
755N	2.02×10^6	5.32×10^5	3.51

Reprinted with permission from ref. 3. Copyright 1990 Marcel Dekker.

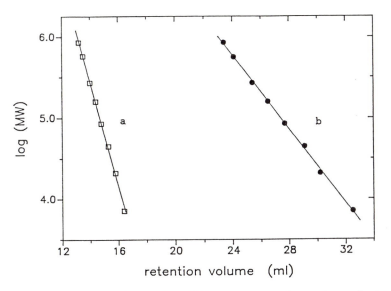

Figure 1   Linear fitted data of log(MW) versus retention volume for PEO standards using: a) SU2 column set with DMF solvent.
b) U4 column set with water/methanol solvent. (Reproduced with permission from ref.4. Copyright 1991 Marcel Dekker).

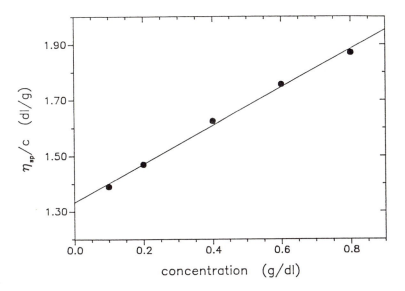

Figure 2   Linear fitted data of reduced viscosity versus concentration for PVPAA 1001, pH 9, 0.2M LiNO$_3$. (Reproduced with permission from ref.4. Copyright 1991 Marcel Dekker).

For nonionic PVPDMAEMAVC and PVPVA with various copolymer compositions studied, Table V shows $\bar{M}_w$ obtained for these copolymers by SEC using the four different column sets with aqueous and organic mobile phases. PVPVA and PVPDMAEMAVC recovery were 100% from all of the columns sets. Also shown in Table V are [η] values for these polymers which show the generally expected increase in [η] with an increase in $\bar{M}_w$ for copolymers of the same monomer compositions. For the copolymers with different monomer compositions, there is not necessarily any correlation between $\bar{M}_w$ and monomer composition.

Table V $\bar{M}_w{}^\dagger$ determined by SEC using aqueous and organic solvents for PVPDMAEMAVC and various copolymer compositions of PVPVA (see Experimental section for descriptions of copolymer compositions, solvents and columns). Intrinsic viscosity of these polymers, at 25°C, in the corresponding solvents used for SEC

	Aqueous Solvent		Organic Solvent			
	Column Set		Column Set			
	U4		SU2	SP2	PP2	
Polymer	$\bar{M}_w$	[η] (dl/g)	$\bar{M}_w$	$\bar{M}_w$	$\bar{M}_w$	[η] (dl/g)
PVPVA:						
E335	28,800	0.265	37,900	36,700	45,000	0.261
E535	36,700	0.363	38,700	38,300	44,500	0.241
E635	38,200	0.330	37,600	37,500	45,100	0.253
E735	56,700	0.429	52,200	52,200	53,800	0.310
I335	12,700	0.176	15,000	16,700	16,000	0.162
I535	19,500	0.222	20,300	22,200	21,600	0.174
I735	22,300	0.261	21,500	24,000	21,400	0.182
W735	27,300	0.265	25,000	27,800	30,600	0.238
S630	51,000	0.424	48,600	49,300	56,000	0.321
PVPDMAEMAVC:						
	82,700	0.620	68,200	73,500	101,000	0.480

$\dagger$ $\bar{M}_w$ relative to PEO standards.
Reprinted with permission from ref. 4. Copyright 1991 Marcel Dekker.

Figure 3 shows SEC traces, overlayed, of PVPVA I series obtained using the SU2 column set with the organic mobile phase. The shapes of these chromatograms are typical of those obtained from the various compositions of PVPVA and PVPDMAEMAVC studied using the SU2 column set. Of the three column sets investigated for the nonionic polymers, SU2, SP2, and PP2, the SU2 column set yielded a separation between the trailing end of the polymer peak and the leading end of the solvent peak whose valley was closest to the baseline.

Because none of these column sets achieved a satisfactory resolution in the low molecular weight range, values of $\bar{M}_n$ were not reproducible, usually with errors greater than 5%. For the copolymers studied, the three column sets all yielded chromatograms with low molecular weight shoulders, which are evident in Figure 3.

Figure 4 shows SEC traces of the PVPVA I series obtained using the U4 column set with the aqueous mobile phase and these chromatograms are typical of those obtained from the various compositions of PVPVA and PVPDMAEMAVC studied. A comparison of the chromatograms obtained using the aqueous mobile

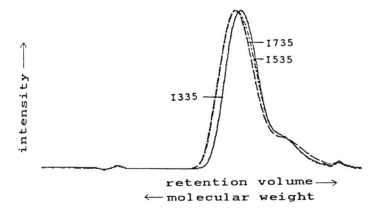

Figure 3 SEC traces of PVPVA, I series, using the SU2 coumn set with DMF solvent. (Reproduced with permission from ref.4. Copyright 1991 Marcel Dekker).

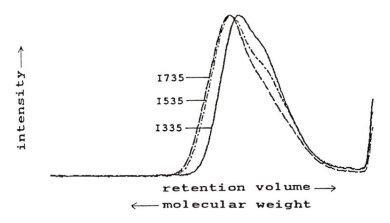

Figure 4 SEC traces of PVPVA, I series, using the U4 column set with water/methanol solvent. (Reproduced with permission from ref.4. Copyright 1991 Marcel Dekker).

phase with chromatograms obtained using the organic mobile phase (Figure 3) with the corresponding column sets, the U4 column set using aqueous mobile phase yielded a separation between the polymer and solvent peak whose valley was closer to the baseline. Therefore, in comparison of SEC results achieved using either the aqueous or organic mobile phase with the corresponding column set, the U4 column set with the aqueous mobile phase yielded the best separation between polymer and solvent peaks. The aqueous mobile phase with the U4 column set, however, also yielded chromatograms with low molecular weight shoulders, similar to the chromatograms obtained using the organic mobile phase with the three two-column sets.

Table VI shows $\bar{M}_w$ obtained by SEC for PVPAA with various copolymer compositions studied using the U4 column set with pH 9 mobile phase. Figure 5 shows SEC traces, overlayed, of the four copolymers. The chromatograms are reasonable but a baseline separation between the low molecular weight end of the polymer peak and the solvent peak could not be achieved. Also shown in Table

Table VI $\bar{M}_w{}^\dagger$ determined by SEC using pH 9 buffer, for various compositions of PVPAA (see Experimental section for descriptions of copolymer compositions and columns). Intrinsic viscosity of these polymers in pH 9 buffer at 25°C

PVPAA	$\bar{M}w$	$[\eta]$ (dl/g)
1005	135,000	1.04
1004	256,000	1.37
1001	318,000	1.33
1030	277,000	----

† $\bar{M}_w$ relative to PEO standards.
Reprinted with permission from ref. 4. Copyright 1991 Marcel Dekker.

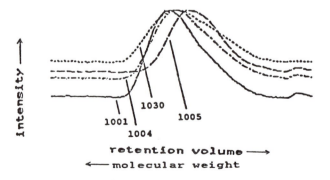

Figure 5 SEC traces of PVPAA copolymers using the U4 column set with pH 9 solvent. (Reproduced with permission from ref.4. Copyright 1991 Marcel Dekker).

VI are $[\eta]$ values which show the generally expected increase in $[\eta]$ with an increase in $\bar{M}_w$. The determination of intrinsic viscosity for PVPAA 1030 was hampered by the apparent insolubility of this polymer in 0.25M NaOH at a concentration greater than 4 g/dl.

Conclusions

Quaternized poly(vinylpyrrolidone-co-dimethylaminoethylmethacrylate), a cationic polymer in a mobile phase of 0.1M TRIS pH 7 buffer with 0.5M LiNO$_3$, was found to elute based on hydrodynamic volume using Waters Ultrahydrogel columns. $\bar{M}_w$ obtained by both SEC/LALLS and SEC with universal calibration were in good agreement.   The most satisfactory SEC results for nonionic poly(vinylpyrrolidone-co-vinylacetate) and poly(vinylpyrrolidone-co-dimethylaminoethylmethacrylate-co-vinylcaprolactam) were achieved using Waters Ultrahydrogel columns with a water/methanol, 0.1M LiNO$_3$ mobile phase.   SEC results for these nonionic copolymers were evaluated qualitatively in terms of resolution between polymer and solvent peaks using columns from Waters, Polymer Laboratories, and Shodex with aqueous and organic mobile phases.   SEC results are reported for anionic poly(vinylpyrrolidone-co-acrylic acid) in 0.1M TRIS pH 9 buffer with 0.1M LiNO$_3$ using Waters Ultrahydrogel columns.

Acknowledgements

The authors wish to thank Edward G. Malawer for helpful advice. The authors are also grateful to John Tancredi and International Specialty Products for permission to publish this work.

Literature Cited

1. Malawer, E.G.; DeVasto, J.K.; Frankoski, S.P; Montana, A.J.
     J. Liq. Chromatogr. **1984**, 7, 441.
2. Senak, L.; Wu, C.S.; Malawer, E.G. *J. Liq. Chromatogr.* **1987**, 10, 1127.
3. Wu, C.S.; Senak, L. *J. Liq. Chromatogr.* **1990**, 13, 851.
4. Wu, C.S.; Curry, J.F.; Senak, L. *J. Liq. Chromatogr.* **1991**, 14, 3331.
5. Wu, C.S.; Senak, L; Malawer, E.G. *J. Liq. Chromatogr.* **1989**, 12, 2901.
6. K.J. Valan *US Patent* 3,914,403 October 21, **1975**.
7. K.J. Valan *US Patent* 3,954,960 May 4, **1976**.
8. Lorenz, D.H.; Murphy, E.J.; Rutherford, J.M.,Jr. US Patent 4,521,404
     June 4, **1985**.
9. Kuo, C.; Provder, T.; Koehler, M.E. Characterization of
     Polymer MWD and Branching with GPC-Viscometry in THF and DMF,
     First International GPC/Viscometry Symposium **1991**, Houston, Tx.
10. Kraemer, E. O. *Ind. Eng. Chem.* **1938**, 20, 1200.
11. Huggins, M. L. *J. Am. Chem. Soc.* **1942**, 64, 2716.

RECEIVED September 9, 1992

Chapter 21

Size-Exclusion Chromatography To Characterize Cotton Fiber

Judy D. Timpa

Southern Regional Research Center, Agricultural Research Service, U.S. Department of Agriculture, P.O. Box 19687, New Orleans, LA 70179

In our laboratory, cotton fibers were dissolved in the solvent N,N-dimethylacetamide with lithium chloride (DMAC/LiCl) and analyzed by size exclusion chromatography (SEC). Cotton fiber (~96% cellulose) has been technically difficult to characterize. This procedure solubilizes fiber cell wall components directly without prior extraction or derivatization, processes that could lead to degradation of high molecular weight components. Molecular weight distributions were determined with DMAC/LiCl as the mobile phase employing commercial SEC columns and instrumentation. A universal calibration was employed by incorporation of viscometer and refractive index detectors. Applications of this technique have focused on elucidating relationships between cotton fiber molecular composition and fiber quality. Primary and secondary wall compositions of cotton fiber have been monitored during fiber development. Cellulose weight average molecular weight correlated with fiber strength determined on fiber classification standards corresponding to a range of fiber lengths and strengths.

As a natural textile material possessing unique properties, cotton fiber has maintained its commercial utilization from earliest civilizations until the present. Cotton fiber is the purest form of naturally-occurring cellulose (94-98%); other higher plants generally contain from 20-50% cellulose. Cellulose is the most abundant biopolymer in the world. Failure to determine the precise molecular properties of cellulose has seriously limited progress in understanding cellulose biosynthesis in plants and effective utilization of cellulosic materials as renewable resources *(1)*. In addition to its commercial importance, the developing cotton fiber has attributes that recommend it as an experimental system of choice for investigation of physiological and biochemical changes accompanying cell elongation and/or maturation *(2)*.

Cellulose, a linear, unbranched polymer, is composed of repeating β-1,4-anhydroglucose units. Procedures to quantitatively measure cellulose content are difficult and time-consuming, and often produce only an approximation of actual cellulose content *(3)*. Characterization of native cellulose is even more difficult than quantitation, since the polymer must be both unchanged in molecular properties and truly representative of the cell wall *(4)*. In the past, lack of suitable nondegrading solvents for underivativized cellulose have hampered the characterization of cotton fiber *(5)*. Recently, wood cellulose solubility in the solvent N,N-dimethylacetamide with lithium chloride (DMAC/LiCl) was optimized for potential commercial applications in generating films and fibers *(6-7)*. The solvent DMAC/LiCl produced homogeneous solutions of cellulose under moderate conditions with little or no degradation in direct contrast to other cellulose solvents that degrade the macromolecular backbone *(6-8)*. Solution of cotton linters was also described. It is well known by cellulose chemists that wood pulps and cotton linters are lower in molecular weight and generally broader in chain length distributions than cotton fiber. The solvent DMAC/LiCl has been used to dissolve proteins, synthetic polyamides, chitin, dextran, amylose, amylopectin, and their derivatives *(8-9)*. A major advantage of the DMAC/LiCl solvent is the opportunity which it provides for the characterization of cellulose starting material and derivatives in solution *(8)*. Methodology for molecular weight characterization employing DMAC/LiCl as the solvent for cellulose from wood pulp was reported employing commercial gel permeation chromatography (GPC) or size exclusion chromatography (SEC) equipment *(10)*.

In our laboratory *(11)*, cotton fibers were dissolved directly in the solvent DMAC/LiCl. This procedure solubilizes fiber cell wall components directly without prior extraction or derivatization, processes that could lead to degradation of high molecular weight components. A size exclusion chromatography (SEC) system employing commercial columns and instrumentation was used with universal calibration *(12)* facilitated by incorporation of viscometer and refractive index detectors *(13)*. It has long been held that cotton fiber mechanical properties (strength and elongation) are affected by the amount of secondary wall *(4,14)*, but measurements of the secondary wall have been difficult. In this report, we present the results obtained for molecular weight distributions (MWD's) for various types of cotton fiber samples dissolved in DMAC/LiCl determined by SEC. Applications of this technique have focused on determination of cotton fiber quality as a function of molecular composition.

Experimental

Sample Preparation. Fiber samples of American Upland cotton (Gossypium hirsutum L.), were dissolved as previously described *(11)*. Ground fiber was added to DMAC (Burdick & Jackson) in a Reacti-Vial (Pierce) in a heating block. Activation was by elevating the temperature to 150°C and maintained at that temperature for 1-2 hr. The temperature was lowered to 100°C followed by addition of dried LiCl. Samples were held at 50°C until dissolved (48 h) and subsequently diluted and filtered. Final concentration of samples was 0.9 to 1.5 mg/mL in DMAC with 0.5% LiCl. At least two dissolutions per sample were made for subsequent SEC analysis.

Chromatography. Filtered cotton solutions were analyzed using a SEC system consisting of an automatic sampler (Waters WISP) with an HPLC pump (Waters Model 590), pulse dampener (Viscotek), viscometer detector (Viscotek Model 100) and refractive index detector (Waters Model 410) *(11)*. The detectors were connected in series. The mobile phase was DMAC/0.5% LiCl pumped at a flow rate of 1.0 mL/min. Columns were Ultrastyragel 10^3, 10^4, 10^5, 10^6 (Waters) preceded by a guard column (Phenogel, linear, Phenomenex). A column heater (Waters Column Temperature System) regulated the temperature of the columns at 80°C. Injection volume was 400 µL with a run time of 65 min. The software package Unical based upon ASYST (Unical, Version 3.02, Viscotek) was used for data acquisition and analysis. Calibration was with polystyrene standards (Toyo Soda Manufacturing) dissolved and run in DMAC/0.5% LiCl. The universal calibration curve was a logarithmic function of the product of the intrinsic viscosity times molecular weight versus retention volume.

Data Plots. Results were plotted with SlideWrite Plus (Advanced Graphics Software) in Figure 4 and with SigmaPlot (Jandel) in Figure 7.

Results and Discussion

Dissolution of Cotton Cellulose. Activation of cellulose is necessary to break inter- and intra-molecular hydrogen bonding in order to achieve dissolution in DMAC/LiCl *(6,7)*. Exposure to vapors of DMAC at 150°C has been the method of choice in our laboratory allowing preparation of samples in a single vial *(11)*. As previously reported, this provides ease of operation for screening large numbers of samples. Cotton fibers were thus directly dissolved in DMAC/LiCl without cleanup or derivatization. A concentrated solution is prepared which is then diluted to be compatible with the mobile phase for SEC analysis. It should be reiterated that for high molecular weight, highly crystalline cellulosic material such as mature cotton fiber, care must be exercised in sample preparation. Sometimes it has been necessary to extend the time of activation, or time of stirring with LiCl, or to shake the solutions for several hours. Direct dissolution of cotton fiber at early stages of development when the cellulose content is lower was successful with the solvent DMAC/LiCl *(21)*. This was not unexpected since DMAC/LiCl has been reported to produce homogenous solutions of a range of natural polymers, proteins and polysaccharides *(6-9)*. For example, solubilization of chitin is easy, and no activation is required as is the case with cellulose *(9)*. There is great advantage to this procedure in avoiding the numerous steps in extraction and derivatization. The potential of the aprotic DMAC/LiCl solvent system appears to be significant. As reported in Dawsey's review *(8)*, despite the number of cellulose solvents available, none offers the capacity for a wide a range of applications and organic reactions including homogeneous solutions under moderate conditions with little or no degradation that DMAC/LiCl does.

SEC Analysis. Ekmanis *(10)* evaluated the effects of column temperature and % LiCl in the mobile phase of DMAC/LiCl for SEC characterization of wood pulp cellulose. He determined that a DMAC/0.5% LiCl mobile phase and an 80°C column temperature would optimize the chromatography. Because of the high viscosity of the cotton fiber solutions, those operating conditions were even more desirable for our analyses. We employed four columns for good separation capability. The wide gap between molecular weights found in primary versus secondary walls of cotton fiber will be discussed subsequently. The universal calibration concept employed by incorporation of dual detectors (refractive index and viscometer) bypasses the need for cellulose standards. There are no cellulose standards available. Polystyrene standards for a wide range of molecular weights dissolved readily in DMAC/0.5% LiCl with no activation necessary.

Applications: Monitoring Cotton Fiber Development. Primary and secondary wall compositions of cotton fiber polymers have been monitored during fiber development by determination of MWD's of DMAC/LiCl solutions *(21)*. Cotton fibers are very long (often >2.5 cm) single cells that differentiate from the epidermal layer of the developing cotton seed. On the day of flowering or anthesis, the cell enters into elongation (primary wall stage) for about 18-21 days. Secondary wall deposition of cellulose chains proceeds for the next 5-6 weeks. The secondary wall contributes the largest amount of cellulose in the composition of the fiber. In Figure 1 is the MWD for cotton fiber from a genetic standard variety (Texas Marker-1, TM-1) sampled at 47 days post anthesis (DPA). By this stage of development even though the boll has not opened, the MWD resembles that of the mature harvested cotton fiber *(see Figure 2, Ref. 11)*. The locations of the primary and secondary walls identified in Figure 1 indicate: (a) lower MW for primary compared to the secondary walls, and (b) the larger weight fraction of material found in the secondary wall. To demonstrate the contrast, the primary wall stage evaluated by sampling fiber at 8 DPA exhibits the MWD shown in Figure 2. The wide separation in levels of DP for the primary wall and secondary wall verify limited previous reports *(1,14)*. However, analytical determination in DMAC/LiCl provide complete polymer profiles with valid MWD's because this procedure solubilizes fiber cell wall components directly without prior extraction or derivatization, processes that could lead to degradation of high molecular weight components.

Analysis of wall polymers of near-isogenic fiber mutants compared to parent fiber was carried out from 10 to 30 DPA covering both the primary and secondary wall stages of development was also carried out in our laboratory. Several near isogenic mutant lines in which specific phases of fiber development are affected have been identified and characterized *(15-16)*. A summary of the fiber characteristics includes: (1) TM-1, parent, wild-type, delta cotton; (2) immature fiber (BTL), secondary wall is not normal; (3) pilose (H2), short, dense plant hairs; (4) Ligon lintless (Li), characterized by distorted plant growth and short fiber (~2mm); and naked seed (N1), no linters (short fuzz fibers) present. Taking a slice through time at 25 DPA is provided in Figure 3 for comparison of two of the mutants and the parent. At 25 DPA in fiber development, elongation (primary wall) stage is complete and secondary wall deposition has increased significantly. Comparisons of the primary and secondary walls can be made by such determinations. These

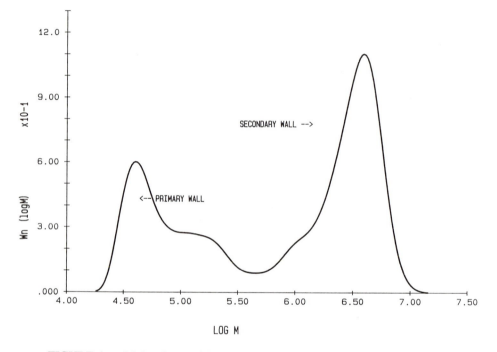

FIGURE 1.   Molecular weight distribution for cotton fiber from Texas
             Marker-1 line (47 days post anthesis) dissolved in
             dimethylacetamide/lithium chloride. Differential distribution
             with weight fraction [Wn(logM)] versus logarithm of
             molecular weight M calculated from universal calibration.

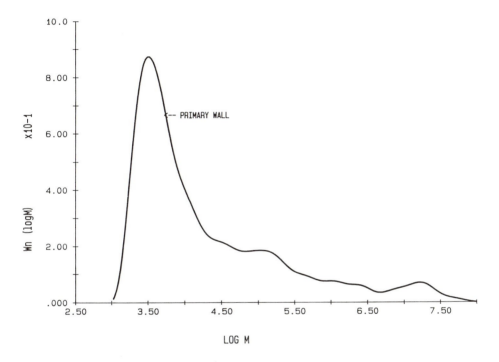

FIGURE 2.   Molecular weight distribution for cotton fiber from Texas Marker-1 line (8 days post anthesis) dissolved in dimethylacetamide/lithium chloride. Primary wall stage of elongation of fiber development. Differential distribution with weight fraction [Wn(logM)] versus logarithm of molecular weight M calculate from universal calibration.

determinations support the concept that secondary wall synthesis of cellulose is of discrete high molecular weight which occurs much earlier than previously observed.

Weight average degrees of polymerization (DP_w's) over the stages of development for the different mutants compared to the parent are plotted in Figure 4. During the early stages when the primary wall is the major contribution to the polymeric composition, the DP_w's were low. With the contribution of secondary wall material increasing, the averages increase. Cellulose from wild type fibers younger than 16 days post-anthesis (DPA) (primary cell wall stages) had lower molecular weights than the cellulose from older fibers (secondary wall stages). Cellulose produced during the secondary wall stages in all of the mutants was identical in molecular weight to the cellulose produced by the wild type. This suggests that the mutants are not defective in their ability to produce cellulose. Two of the mutants started producing high molecular weight cellulose several days earlier than the wild type, indicating some alteration in the normal developmental switch from elongation growth to secondary wall synthesis. This information aids research applications aimed at biotechnological improvement of cotton fiber strength.

Applications: Relating Molecular Weight to Cotton Fiber Strength. Cotton fiber quality is determined by length, strength and fineness. We are evaluating the molecular composition as a function of the variety and environment of growth with respect to development of fiber quality. Exploratory study of the influence of water supply during growth of the cotton plant was carried out *(17)*. Data showed that cellulose molecular weight for the secondary wall was characteristic for a given genotype and may be significantly changed with the environmental conditions under which the cotton was grown. The ranking of three high strength cotton fibers samples correlated with the ranking by cellulose molecular weight *(18)*.

Fiber classification standards with a range of lengths and strengths (H.H. Ramey, personal communication) were sampled. The two extremes of samples are shown by plots of the MWD's in Figures 5 and 6, respectively. The shortest length fiber (0.903 in) with a Stelometer strength of 21.4 g/tex had secondary wall composition $DP_w = 15000$ (Figure 5). The longer fiber (1.236 in) with much higher strength (31.0 g/tex) had $DP_w = 23700$ (Figure 6). It is interesting that the longer, higher strength fiber had multiple peaks present both in the secondary and primary wall regions. It should also be noted that the primary wall peaks for the longer, stronger fiber are at higher molecular weight. It was not known if this multiplicity represented sampling differences with the bales. The relation of the average DP_w's to strength is shown in Figure 7. The general correlation is evident but molecular compositional profiles indicate correlation of higher average molecular weights with greater length and strength within the series. However, comparison of molecular composition profiles of the polymer comprising the cotton fiber demonstrated that populations of cellulose chains can vary to produce similar fiber strength values. These results led us to pursue ongoing assessments of fibers with more detailed varietal and growth history information.

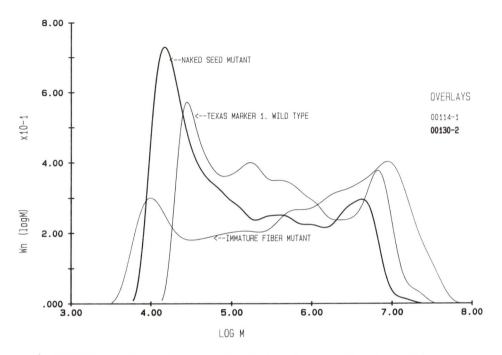

FIGURE 3.   Molecular weight distributions for cotton fiber from wild-type and two mutants at 25 days post anthesis. Differential distribution with weight fraction [Wn(logM)] versus logarithm of molecular weight M calculate from universal calibration.

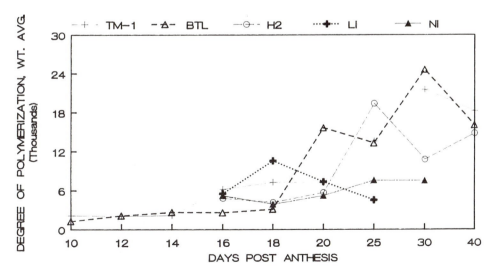

FIGURE 4.   Weight average degrees of polymerization for cotton fiber samples for TM-1 and four mutants over the stages of development. TM-1 = Texas Marker-1; BTL = immature fiber; H2 = pilose, Li = Ligon lintless; N1 = naked seed.

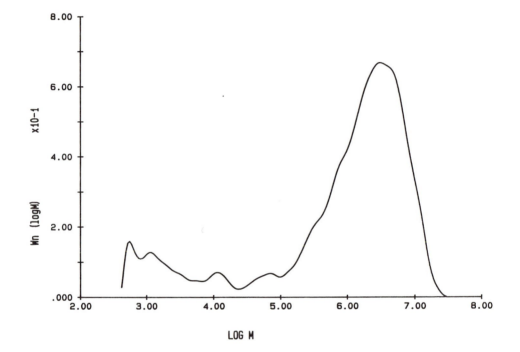

FIGURE 5.   Molecular weight distribution for mature, field grown, mechanically harvested cotton fiber with low strength. Fiber physical properties: length =0.903 in; strength = 21.4 g/tex; micronaire = 4.22; crystallinity = 93.9%.

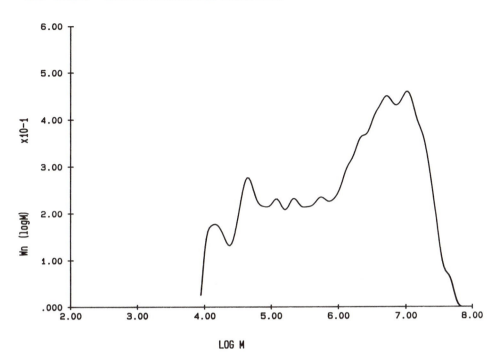

FIGURE 6.   Molecular weight distribution for mature, field grown, mechanically harvested cotton fiber with high strength. Fiber physical properties: length = 1.236 in; strength = 31.0 g/tex; micronaire= 3.7; crystallinity = 93.5%.

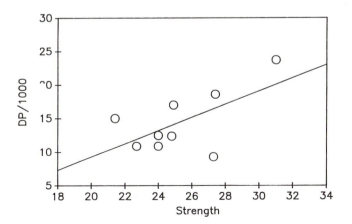

FIGURE 7.   Relationship between weight average degree of polymerization and strength for cotton fibers with a range of lengths and strengths.   Line shown is least squares fit with R value of 0.62.

Applications: Characterization of Other Polysaccharides (Starch). Extension of this technique to other polysaccharides has been undertaken to address processing variables in the food industry. Extrusion-induced fragmentation of starch was measured by SEC *(20)*. Corn meal and two extrudates were dissolved in DMAC/LiCl using the same procedure developed for cotton fiber and MWD's were determined. The automated SEC procedure represents a significant advance in rapid quantitative assessment of starch fragmentation in extruded food products.

Summary

Direct dissolution of cotton fibers in DMAC/LiCL with subsequent SEC analysis to obtain MWD's provides a valuable tool for examination of molecular compositional changes during fiber development and subsequent processing. Correlations with physical/mechanical properties, particularly strength, are being established. We are testing the hypothesis that cotton fiber quality is intricately linked with the biosynthesis of cell wall polymers, specifically cellulose. Determination of molecular variability as a function of genotype and growth conditions can be employed to improve selection for varieties to obtain optimized products for textile processing. Improvements in cellulosic and other natural products can only come by acquiring better basic understanding of the product and processes of biosynthesis.

Safety Considerations

N,N-Dimethylacetamide is an exceptional contact hazard that may be harmful if inhaled or absorbed through skin and may be fatal to embryonic life in pregnant females (Baker Chemical C. N,N-Dimethylacetamide, Material Safety Data Sheet, 1985, D5784-01; pp. 1-4).

Acknowledgments

The author would like to thank the following for their cooperation: B. A. Triplett, H. H. Ramey, B. Wasserman, H. Zeronian, W. R. Meredith.

Literature Cited

1. Marx-Figini, M. In *Cellulose and Other Natural Polymer Systems: Biogenesis, Structure, and Degradation;* Brown, R. M. Ed.; Plenum Press, New York, 1982, pp. 243-271.
2. Basra, A. S.; Malik, C. P. *Int. Rev. Cytol.* 1984, *89*, 65-113.
3. Ladisch, C. M. *New Methods of Cellulose Analysis in Selected Materials Based on Total Hydrolysis to Sugars.* Purdue University, Ph.D. Thesis 1978.
4. Goring, D. A. I.; Timell, T. E. *TAPPI* 1962, *45*, 454-460.
5. Delmer D. L. *Ann. Rev. Plant Physiol.* 1987, *38*, 259-290.
6. Turbak, A. B. In *Agricultural Residues;* Soltes, E.J. Ed. ACS, Vol XIII, Academic, New York, NY, 1983, pp. 87-99.

7.   McCormick, C. L.; Callais, P. A.; Hutchinson, B. H. *Macromolecules* 1985, *18*, 2394 -2401.
8.   Dawsey, T. R.; McCormick, C. L. *Rev. Macromol. Chem. Phys.* 1990, *C30*, 403-440.
9.   Terbojevich, M.; Carraro, C.; Cosani, A.; Marsano, E. *Carbohydr. Res.* 1988, *180*, 73-86.
10. Ekmanis, J. L. *Am. Lab. News*, 1987, Jan/Feb, 10.
11. Timpa, J. D. *J. Agri. Food Chem.* 1991, *39*, 270-275.
12. Grubisic, A.; Rempp, P.; Benoit, H. A. *Polym. Lett.* 1967, *5*, 753-759.
13. Haney, M. A. *Am. Lab.* 1985, *17*, 116-126.
14. Hessler, L. E.; Merola, G. V.; Berkley, E. E. *Text. Res. J.* 1948, *18*, 628-634.
15. Triplett, B.A., Busch, W.H., Goynes, W.R. *In Vitro, Cell. & Develop. Biol.* 1989, *25*, 197-200.
16. Triplett, B.A. *Text. Res. J.* 1990, *60*, 143-148.
17. Timpa, J. D.; Wanjura, D.F. In *Cellulose and Wood: Chemistry and Technology*; Schuerch, C. Ed., John Wiley, New York, NY. 1989, pp. 1145-1155.
18. Timpa, J. D.; Ramey, H. H. *Textile Res. J.* 1989, *59*, 661.
19. Timpa, J. D. In *Proc. Beltwide Cotton Production Research Conf. 1990: Cotton Physiology,* National Cotton Council, Memphis, TN, 1990, p. 626.
20. Wasserman, B. P.; Timpa, J. D. *Starch/Starke* 1991, *43*, 389-392.
21. Timpa, J. D.; Triplett, B. A. *Planta*, in press.

RECEIVED September 28, 1992

Chapter 22

Structural Analysis of Aggregated Polysaccharides by High-Performance Size-Exclusion Chromatography–Viscometry

Marshall L. Fishman, David T. Gillespie, and Branka Levaj

Eastern Regional Research Center, Agricultural Research Service, U.S. Department of Agriculture, 600 East Mermaid Lane, Philadelphia, PA 19118

Three high performance size exclusion columns placed in series were calibrated in terms of radii of gyration (R_g) and hydrodynamic volume (intrinsic viscosity ([IV]) x molecular weight (M)) with a series of pullulan and dextran standards ranging in M from 853,000 to 10,000. Online detection was by differential refractive index (DRI) and viscometry (DP). Two forms of universal calibration were employed to obtain R_g, [IV], and M for pectin, a class of complex plant cell-wall polysaccharides. For pectins extracted from a large number of sources, a Mark-Houwink plot (log [IV] vs. log M) gave a correlation coefficient of 0.2; whereas, a plot of log [IV] against log R_g gave a correlation coefficient of 0.9. These results in addition to those from the analysis of several pectins from peach fruit indicated that pectins were aggregated and highly asymmetric in shape. The scaling law exponents for pectins were effected by both shape and state of aggregation, rather than shape alone.

The tendency to aggregate significantly complicates the structural analysis of many polysaccharides (1). Pectin is an example of aggregated polysaccharides (2,3), consisting of a group of closely related anionic polysaccharides found in the cell walls of all higher plants (4). Recently, we have shown that pectin from a variety of sources is an associated colloid comprised of five macromolecular-sized species when analysed by HPSEC (5,6). By employing HPSEC/viscometry with curve fitting of the chromatograms and two forms of universal calibration (7), we obtained the root mean square (rms) radii of gyration (R_g), intrinsic viscosities ([IV]), and molecular weights (M) of the five components and their global averages. Here, we examine the effect of aggregation on the inter-relationships between [IV], R_g, and M.

Experimental

Materials. Pectins, referred to as "by-product" pectins, were extracted from beet pulp, the peels of mangoes, oranges, mandarin oranges, grapefruits, pomegranates, and artichokes, the skin of garlic and peas, carrot and colocasia wastes, and garlic foliage. Typically the pectin source was extracted for 1 h at 90°C with 0.5 % w/v ammonium oxalate solution, precipitated with acidified alcohol, and dried (8). Then, 1g of the dried pectin was dissolved in 60 mL of 0.01 M sodium phosphate buffer containing 0.01M EDTA (pH 6.05), stirred overnight at 4°C, dialyzed against 4 changes of water over 48 h, centrifuged at 32,000g for 1 h at 5°C to remove trace insolubles, and lyophilized.

Peach pectins were obtained from the mesocarp of hard, melting flesh (MF) "Redskin" peaches and from non-melting flesh (NMF) "Suncling" peaches. Chelate soluble (CSP) and mildly alkaline soluble pectin (ASP) were extracted sequentially from isolated, washed cell walls according to a procedure described by Gross (9).

Chromatography. High performance size exclusion was performed as reported earlier (10). Pectin, dissolved in 0.05 M $NaNO_3$ or NaCl, was passed through a 0.4 μM Nucleopore filter and equilibrated overnight at 35°C in capped bottles prior to chromatography. The mobile phase was either 0.05 $NaNO_3$ or NaCl in HPLC grade water. Solvent was degassed prior to connecting to the system and inline with a model ERC 3120 degasser, Erma Optical Co., Tokyo. The solvent delivery system was a model 334, Beckman Instr., Palo Alto, CA. The nominal flow rate was 0.5 mL/min. The pumping system was fitted with a Beckman pulse filter and two model M45 pulse dampners, Waters Assoc, Millford, MA, mounted on a plate and separated by 15 ft of coiled capillary tubing (i.d., 0.01 inches). Sample injection was with a Beckman model 210 valve. The injected sample volume was 100 μL. Three columns were employed in series, a Micro-Bondagel E-High A, E-1000, Waters Assoc.(300x3.9 mm) and a Synchropak GPC-100 (250x4.6 mm) Synchrom, Inc., Linden, IN. The viscosity detector (differential pressure detector, DP) was a model 100 differential viscometer, Viscotek Corp., Porter, TX or an inhouse single coil model described in reference 10. When viscosity detection was with the model 100, injected sample concentrations ranged from 0.53 to 0.57 mg/mL (i.e., peach pectins) whereas viscosity detection with the inhouse model required sample concentration in the range 2.5 to 2.7 mg/mL ("by-product" pectins). Differential refractive index (DRI) was measured with a model ERC 7810 monitor, Erma Optical Co., Tokyo. Chromatography columns were thermostated in a temperature controlled water bath at 35 ± 0.003°C and the cells of the refractive index and viscosity monitors were thermostated also at 35°C. Data collection and flow rate measurement have been described previously (11).

Curve fitting. The partially resolved, overlapped components of the DRI and the DP detector chromatograms were determined with the aid of ABACUS, version D.2, a nonlinear least-squares curve fitting program. The dead volume between the DRI and DP detectors was measured as 125 ± 1 μL by matching the front end of chromatograms from a narrow P-50 pullulan standard normalized for area from the respective detectors. For fitting of DP traces, values of component peak position, quarter width at half height (sigma) and number of Gaussian peak components were obtained from DRI traces as already described (5). Since the two detectors only differed in their sensitivity of response towards pectin, only peak heights were iterated until the sum of the squares of the point by point residuals between the calculated curve reconstructed from the components and the experimental trace were minimized to convergence.

Column Calibration. As previously described congruent calibration curves were obtained by plotting log R_g or log [IV]M against column partition coefficient,(K_{av}), for a series of narrow molecular weight distribution (MWD) pullulan and broad MWD dextran standards (i.e., dextrans with polydispersities ranging from 1.39 to 2.91) (5,7). These calibration curves were used to obtain R_g and M for unknown pectin samples by the "universal calibration" procedure. According to this procedure, pectins will co-elute with dextrans and pullulans which have identical values of either R_g or product of [IV] and M. Values of M or R_g for pectins were calculated by transforming partition coefficients along the pectin refractive index response to R_g or to the product [IV]xM. Transformations were obtained from the dextran-pullulan calibration curves (11). To obtain M as a function of K_{av}, the product of [IV]M as a function of K_{av} was divided by [IV] which also was obtained as a function of K_{av} by the online viscosity detector. In cases where component analysis was possible, the molecular weight or radius of gyration for the component was obtained from the peak maximum of the component. The weight fraction of the component was obtained from the component area under the refractive index trace. Weight average properties were obtained by summing over the components as described previously (7). In cases where component analysis was not possible, continuum calculations were carried out. In these cases, corrections for bandspreading were made by the GPCV2 procedure (12).

Results and Discussion

For a macromolecule dissolved in a good solvent at constant temperature, [IV], R_g, and M are interrelated through equation 1, the modified Einstein equation (13).

$$[IV] = A(R_g)^3 / M \tag{1}$$

In the case of a single, linear polymer chain, an increase in the degree of polymerization will result in increases in intrinsic viscosity ([IV]), radius of gyration (R_g), and molecular weight (M), in such a manner as to maintain a constant value of A, the proportionality constant in equation 1. In the case of pectin, a highly asymmetric, aggregated, polyelectrolyte (2), the question arises as to whether some circumstances might exist in which [IV] and M, are not dependent on R_g. A Mark-Houwink plot was constructed for the 12 "by-product" pectins by plotting log [IV] against log M_w (weight average molecular weight) (Figure 1). M_w was calculated from five curve-fitted components of the HPSEC chromatograms as described previously (7). The correlation coefficient for this data was 0.2 whereas a plot of log [IV] against log R_{gw} (weight average radius of gyration) (Figure 2) gave a correlation coefficient of 0.9. The Mark-Houwink scaling law exponent was 0.38 which is the value expected for a macromolecule more compact than a random coil in an ideal solvent (14). A value of 0.89 was found for the scaling law exponent, x, relating R_g and M_w, which is obtained from equation 2 (10).

$$[IV] = C(R_g)^{3 - (1/x)} \tag{2}$$

The value of x is close to the expected value for a rigid rod (15).

The finding that [IV] was more highly correlated with R_g than it was with M was tested further by measuring [IV], R_g, and M for pectins from two solubility fractions in each of two varieties of peaches. In the case of the chelate soluble pectin (CSP) fractions unlike the alkaline soluble pectin (ASP) fractions, differential refractive index (DRI) and differential pressure (DP) chromatograms could not be fitted with the same set of components (cf. ASP and CSP chromatograms from (non melting flesh) peaches in Figures 3 and 4). Thus the four pectins were compared with weight average global parameters rather than parameters for the components. As indicated by the data in Table I, for the two varieties, R_g and [IV] are substantially higher in the CSP fraction than in the ASP fraction whereas M differs much less between the two fractions and does not appear to be highly correlated with [IV].

Previously (5), we have shown that the larger pectin components can be dissociated into smaller ones by dialysing against 0.05 M NaCl. In the course of three separate but similar extractions of ASP from melting flesh peaches, the largest component of sample 3 appears to have undergone dissociation during extraction. The results from these experiments are found in Table II. For the largest component in all three extractions, R_g was about the same, 42.2 ± 1.9 nm whereas there were appreciable differences in [IV] and M for samples 1 and 3, ranging from 5.7 ± 0.3 dL/g and 316 ± 29 x 10^3 to 3.62 ± 0.01 dL/g and 458 ± 13. In accordance with equation 1, at constant R_g, [IV] was inversely related with M. Such behavior would be possible under circumstances in which R_g and M were independent variables. One example would be two highly asymmetric molecules with identical contour lengths but differing in thickness, e.g., two aggregated rods which differed in degree of aggregation but not in length. For rod-like molecules, R_g and M are dependent variables when length changes but independent variables when only thickness changes. Furthermore increases in molecular weight which are only related to increases in thickness decrease viscosity whereas increases in molecular weight which are only related to increases in length increase viscosity.

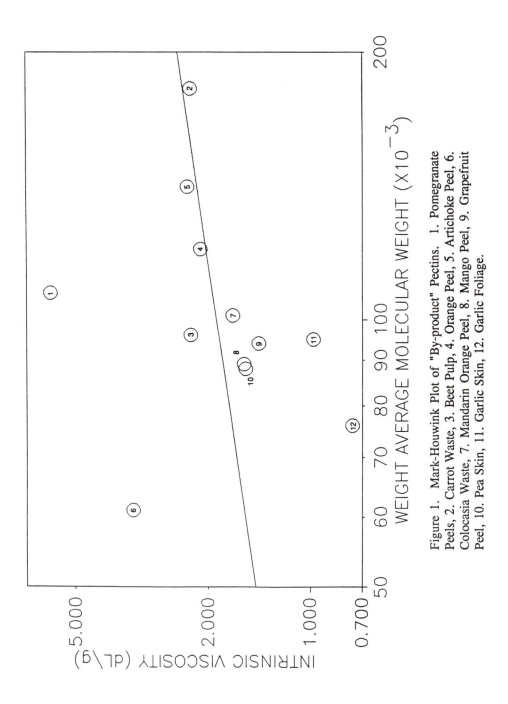

Figure 1. Mark-Houwink Plot of "By-product" Pectins. 1. Pomegranate Peels, 2. Carrot Waste, 3. Beet Pulp, 4. Orange Peel, 5. Artichoke Peel, 6. Colocasia Waste, 7. Mandarin Orange Peel, 8. Mango Peel, 9. Grapefruit Peel, 10. Pea Skin, 11. Garlic Skin, 12. Garlic Foliage.

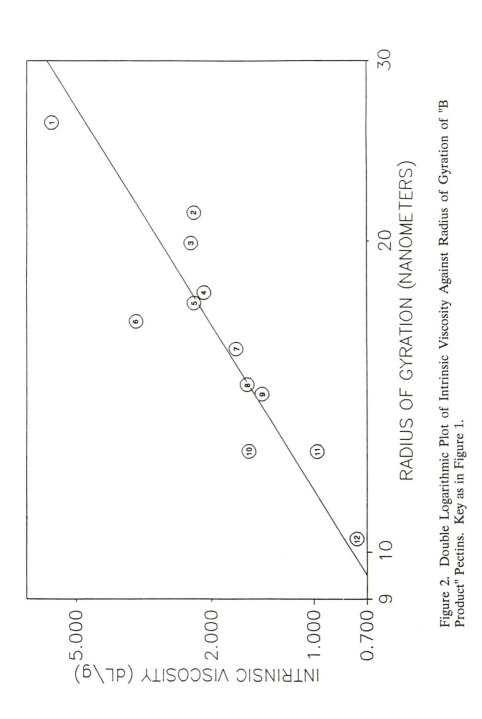

Figure 2. Double Logarithmic Plot of Intrinsic Viscosity Against Radius of Gyration of "B Product" Pectins. Key as in Figure 1.

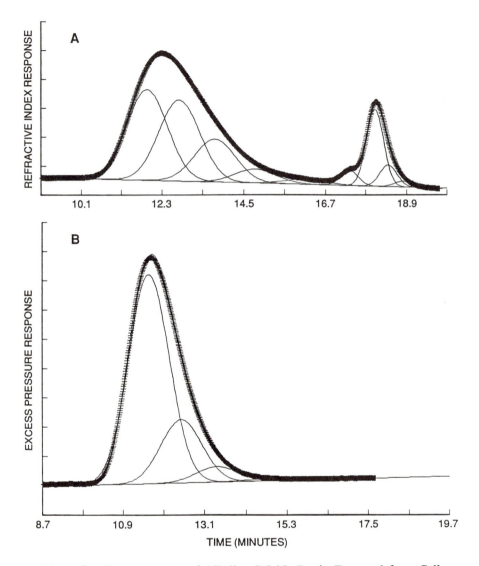

Figure 3. Chromatograms of Alkaline Soluble Pectin Extracted from Cell Walls of "Non Melting Flesh Peaches". Mobile phase, 0.05 M NaNO$_3$; nominal flow rate, 0.5 mL/min.; injection volume, 100 μL; Injected concentration 0.55 mg/mL. Thick line, experimental; thin line, calculated detection. Macromolecular components referred to in text are numbered 1-5, left to right. (A) Detector, differential refractive index; (B) detector, differential pressure (differential viscosity).

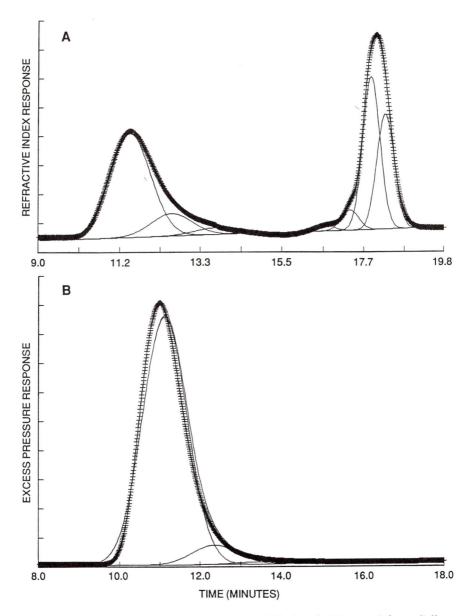

Figure 4. Chromatograms of Chelate Soluble Pectin Extracted from Cell Walls of "Non Melting Flesh Peaches". Mobile phase, 0.05 M $NaNO_3$; nominal flow rate, 0.5 mL/min.; injection volume, 100 μL; Injected concentration 0.53 mg/mL. Thick line, experimental; thin line, calculated detection. Macromolecular components referred to in text are numbered 1-5, left to right. (A) Detector, differential refractive index; (B) detector, differential pressure (differential viscosity).

Table I. Comparison of Weight Average Properties of Pectins
from Two Varieties of Peaches

Fraction	CSP			ASP		
Variety	Rg[1]	[IV][2]	Mx10^{-3}	Rg	[IV]	Mx10^{-3}
NMF[3,4]	49.4 ± 0.9	11.8 ± 0.3	206 ± 4	25.5 ± 2.0	3.3 ± 0.5	159 ± 12
MF[5]	40.6 ± 0.9	12.2 ± 0.3	125 ± 23	27.0 ± 2.0	3.0 ± 0.5	204 ± 16

[1]Nanometers   [2]dL/g   [3]non melting flesh   [4]$\pm$S.D. of 3 measurements   [5]melting flesh

Table II. Properties for Components from ASP "Melting Flesh" Pectin

Component	1	2	3	4	5
Weight Fract.[1]					
Sample 1	0.408 ± 0.00	0.356 ± 0.00	0.184 ± 0.00	0.052 ± 0.00	-----------
Sample 2	0.365 ± 0.00	0.350 ± 0.00	0.200 ± 0.00	0.058 ± 0.00	0.027 ± 0.002
Sample 3	0.346 ± 0.00	0.347 ± 0.00	0.227 ± 0.00	0.066 ± 0.00	0.015 ± 0.002
R_g (nm)					
Sample 1	43.4 ± 1.3	22.3 ± 0.5	12.7 ± 0.3	7.8 ± 0.2	-----------
Sample 2	40.4 ± 0.3	20.3 ± 0.1	11.3 ± 0.1	6.5 ± 0.1	2.9 ± 0.1
Sample 3	41.9 ± 0.5	20.2 ± 0.3	11.4 ± 0.2	6.9 ± 0.3	3.3 ± 0.2
[IV] (dL/g)					
Sample 1	5.7 ± 0.3	2.31 ± 0.05	1.08 ± 0.09	0.43 ± 0.11	-----------
Sample 2	5.2 ± 0.1	2.32 ± 0.02	0.85 ± 0.03	0.55 ± 0.01	0.55 ± 0.07
Sample 3	3.62 ± 0.01	2.43 ± 0.03	0.63 ± 0.03	0.73 ± 0.04	0.95 ± 0.18
% Sp V[2]					
Sample 1	69.0 ± 1.3	24.5 ± 1.4	5.9 ± 0.3	0.7 ± 0.2	-----------
Sample 2	64.8 ± 0.2	27.7 ± 0.2	5.8 ± 0.2	1.1 ± 0.1	0.5 ± 0.1
Sample 3	54.4 ± 0.4	36.6 ± 0.1	6.3 ± 0.3	2.1 ± 0.1	0.6 ± 0.2
M x 10^{-3}					
Sample 1	316 ± 29	155 ± 11	71 ± 5	45 ± 17	-----------
Sample 2	291 ± 5	122 ± 2	65 ± 2	19 ± 1	0.18 ± 0.02
Sample 3	458 ± 13	111 ± 2	90 ± 6	17 ± 3	0.15 ± 0.06

[1]Sample 1 average $\pm$ S.D. of 5 measurements; samples 2 and 3 average $\pm$ S.D. of 3 measurements.   [2]Percentage specific viscosity.

Table III contains [IV], R_g and molecular weight values for components of two groups of "by-product" pectin. Each of these groups was chosen because their components had R_g values that were closely similar. Component 1 of beet, orange, and carrot pectin; and components 1 and 2 of pea skin, grapefruit and garlic skin pectin had [IV]'s which were different. As with the peach components, there is an inverse relationship between [IV] and molecular weight for these components.

Conclusions

These results are consistent with the hypothesis that pectin is comprised of aggregated rods, aggregated segmented rods or a combination of both. The low correlation of log [IV] and log M and the relatively higher correlation of log [IV] and log R_g is consistent with the finding that pectins of comparable R_g have [IV]'s and M's which are inversely related, if they are aggregated to different extents. Furthermore, such a situation is consistent with the modified Einstein Law relating [IV], M, and R_g.

In this work, we have produced evidence that asymmetric molecules which are aggregated to different extents but have identical R_g values will co-elute on a size exclusion column in spite of differing [IV] and M values. An important consequence of this finding is that for these kinds of macromolecules, e.g., pectins, universal calibration rather than calibration of the column by pectins of "known" molecular weight could be a better procedure for determining molecular weights by HPSEC.

Polysaccharides are ubiquitously distributed throughout the world of plants, animals and microorganisms (16). They are important industrially and in biological processes. Although they are involved in a variety of roles, many details remain to be learned at the molecular level concerning structural-functional relationships between polysaccharides and the complex systems in which they exist. As an example, in the case of pectin, the work in this report and several others (5-7,10,17) indicates the existence of pectin quaternary structure. We believe that a better knowledge of pectins' quaternary structure and under what conditions it changes is extremely important in understanding the mechanisms by which pectin functions as a dietary fiber which lowers blood cholesterol and reduces glucose intolerance in diabetics, contributes to the texture of fruits, vegetables, and their processed products, acts as a chemical messenger to defend plants against attack by pathogens and induces metabolic processes important in plant growth, development and senescence (18). A better understanding of these mechanisms would supply information which could aid in delaying heart failure, reducing the incidence of certain cancers through proper nutrition; and aid in the development of more disease-resistant plants whose edible products would taste better, would be less susceptible to post harvest deterioration, and would be more readily processed.

Table III. Properties of Components from "By-Product" Pectins

Component	1	2	3	4	5
$R_g{}^1$ (nm)	39.7±1.0	20.7±1.0	10.9±0.7	5.4±0.4	2.6±0.2
[IV] (dL/g)					
Beet	4.4±0.1	2.7±0.1	0.94±0.04	0.90±0.23	1.68±0.24
Orange	4.0±0.1	2.9±0.1	0.94±0.04	0.89±0.16	0.82±0.09
Carrot	3.6±0.1	2.7±0.1	1.07±0.05	0.99±0.25	1.06±0.49
M x 10^{-3}					
Beet	383±33	120±12	60±9	9.7±3.8	0.5±0.1
Orange	404±8	95±6	48±5	7.0±2.3	0.8±0.1
Carrot	475±21	125±11	42±4	6.8±2.3	1.1±0.5
% Sp V					
Beet	40.1±1.1	41.3±0.4	13.0±1.0	4.2±0.4	1.4±0.3
Orange	33.5±1.6	44.7±0.6	16.2±0.5	4.5±0.7	1.1±0.1
Carrot	43.5±1.7	37.6±1.2	13.8±1.4	3.9±0.5	1.3±0.7
$R_g{}^1$ (nm)	34.0±0.8	17.6±0.7	9.7±0.4	5.5±0.2	2.8±0.2
[IV] (dL/g)					
Pea Skin	4.3±0.1	2.0±0.1	0.67±0.01	0.38±0.08	0.22±0.12
Grapefruit	3.3±0.2	2.3±0.1	0.78±0.06	0.36±0.14	0.74±0.36
Garlic Skin	2.5±0.2	1.7±0.1	0.65±0.03	0.27±0.03	0.38±0.42
M x 10^{-3}					
Pea Skin	237±5	103±5	51±8	13±5	4.0±2
Grapefruit	362±33	96±7	54±2	23±9	1.4±0.6
Garlic Skin	444±22	111±10	52±2	26±3	6.4±4.7
% Sp V					
Pea Skin	37.5±0.9	43.1±0.3	16.1±0.9	3.1±0.1	0.5±0.4
Grapefruit	26.9±2.6	43.4±1.4	24.3±2.8	4.0±1.6	1.4±0.9
Garlic Skin	23.5±0.9	40.8±0.8	28.4±1.2	6.2±1.2	1.1±1.0

[1]Average ± S.D. of 3 pectins x 3 replicates = 9 measurements; [IV], M and % SP V (percentage specific viscosity) are average ± S.D. of 3 measurements.

Literature Cited

1. Morris, E.R.; Norton, I.T. In *Polysaccharide Aggregation in Solution and Gels*; Wyn-Jones, E.; Gormally, J. Eds.; Studies in Theoretical and Physical Chemistry - 26; Elsevier Scientific Publishing Company, 1983; Ch. 19.
2. Fishman, M.L.; Pepper, L.; Pfeffer, P.E. In *Water Soluble Polymers— Beauty with Performance*; Glass, J.E. Ed.; Advances in Chemistry Series - 213; American Chemical Society, 1986; Ch. 3.
3. Jordan, R.C.; Brant, D.A. *Biopolymers* **1978**, *17*, 2885.
4. Jarvis, M.C. *Plant Cell Environ.* **1984**, *7*, 153.
5. Fishman, M.L.; Gross, K.C.; Gillespie, D.T.; Sondey, S.M. *Arch. Biochem. Biophys.* **1989**, *274*, 179.
6. Fishman, M.L.; El-Atawy, Y.S.; Sondey, S.M.; Gillespie, D.T.; Hicks, K.B. *Carbohydr. Polym.* **1991**, *15*, 89.
7. Fishman, M.L.; Gillespie, D.T.; Sondey, S.M.; El-Atawy, Y.S. *Carbohydr. Res.* **1991**, *215*, 91.
8. El-Atawy, Y.S. Study on Pectin Production from Some By-products of Food Factories. PhD. Thesis; Faculty of Agriculture (Moshtohar); Zagazig University, 1984.
9. Gross, K.C. *Physiol. Plant.* **1984**, *62*, 25.
10. Fishman, M.L.; Gillespie, D.T.; Sondey, S.M.; Barford, R.A. *J. Agric. Food Chem.* **1989**, *37*, 584.
11. Fishman, M.L.; Damert, W.C.; Phillips, J.G.; Barford, R.A. *Carbohydr. Res.* **1987**, *160*, 215.
12. Yau, W.W.; Kirkland, J.J.; BLY, D.D. In *Modern Size Exclusion Chromatography*; Wiley: New York, 1979 p. 301.
13. Flory, P.J. In *Principles of Polymer Chemistry*; Cornell University Press: Ithaca, N.Y., 1953; p. 611.
14. Tanford, C. In *Physical Chemistry of Macromolecules*; Wiley: New York, 1961; p. 412.
15. Yau, W.W.; Kirkland, J.J; Bly, D.D. In *Modern Size Exclusion Liquid Chromatography*; Wiley: New York, 1979; p. 44.
16. Rees, D.A. In *Polysaccharide Shapes*: Wiley, New York, 1977.
17. Fishman, M.L.; Cooke, P.; Levaj, B.; Gillespie, D.T.; Sondey, S.M. and Scorza, R. *Arch. Biochem. Biophys.* **1992**, *294*, In press.
18. Fishman, M.L. *ISI Atlas of Science: Biochemistry*, **1988**, *1*, 215.

RECEIVED September 30, 1992

INDEXES

Author Index

Affiliation Index

Subject Index

Production: Betsy Kulamer
Indexing: Deborah H. Steiner
Acquisition: Rhonda Bitterli

Printed and bound by Maple Press, York, PA

Bestsellers from ACS Books

The ACS Style Guide: A Manual for Authors and Editors
Edited by Janet S. Dodd
264 pp; clothbound ISBN 0–8412–0917–0; paperback ISBN 0–8412–0943–X

The Basics of Technical Communicating
By B. Edward Cain
ACS Professional Reference Book; 198 pp;
clothbound ISBN 0–8412–1451–4; paperback ISBN 0–8412–1452–2

Chemical Activities (student and teacher editions)
By Christie L. Borgford and Lee R. Summerlin
330 pp; spiralbound ISBN 0–8412–1417–4; teacher ed. ISBN 0–8412–1416–6

Chemical Demonstrations: A Sourcebook for Teachers,
Volumes 1 and 2, Second Edition
Volume 1 by Lee R. Summerlin and James L. Ealy, Jr.;
Vol. 1, 198 pp; spiralbound ISBN 0–8412–1481–6;
Volume 2 by Lee R. Summerlin, Christie L. Borgford, and Julie B. Ealy
Vol. 2, 234 pp; spiralbound ISBN 0–8412–1535–9

Chemistry and Crime: From Sherlock Holmes to Today's Courtroom
Edited by Samuel M. Gerber
135 pp; clothbound ISBN 0–8412–0784–4; paperback ISBN 0–8412–0785–2

Writing the Laboratory Notebook
By Howard M. Kanare
145 pp; clothbound ISBN 0–8412–0906–5; paperback ISBN 0–8412–0933–2

Developing a Chemical Hygiene Plan
By Jay A. Young, Warren K. Kingsley, and George H. Wahl, Jr.
paperback ISBN 0–8412–1876–5

Introduction to Microwave Sample Preparation: Theory and Practice
Edited by H. M. Kingston and Lois B. Jassie
263 pp; clothbound ISBN 0–8412–1450–6

Principles of Environmental Sampling
Edited by Lawrence H. Keith
ACS Professional Reference Book; 458 pp;
clothbound ISBN 0–8412–1173–6; paperback ISBN 0–8412–1437–9

Biotechnology and Materials Science: Chemistry for the Future
Edited by Mary L. Good (Jacqueline K. Barton, Associate Editor)
135 pp; clothbound ISBN 0–8412–1472–7; paperback ISBN 0–8412–1473–5

For further information and a free catalog of ACS books, contact:
American Chemical Society
Distribution Office, Department 225
1155 16th Street, NW, Washington, DC 20036
Telephone 800–227–5558